Ruth Hindshaw

Chemical weathering and calcium isotope fractionation

Ruth Hindshaw

Chemical weathering and calcium isotope fractionation

Damma Glacier, Switzerland

Südwestdeutscher Verlag für Hochschulschriften

Impressum/Imprint (nur für Deutschland/only for Germany)
Bibliografische Information der Deutschen Nationalbibliothek: Die Deutsche Nationalbibliothek verzeichnet diese Publikation in der Deutschen Nationalbibliografie; detaillierte bibliografische Daten sind im Internet über http://dnb.d-nb.de abrufbar.
Alle in diesem Buch genannten Marken und Produktnamen unterliegen warenzeichen-, marken- oder patentrechtlichem Schutz bzw. sind Warenzeichen oder eingetragene Warenzeichen der jeweiligen Inhaber. Die Wiedergabe von Marken, Produktnamen, Gebrauchsnamen, Handelsnamen, Warenbezeichnungen u.s.w. in diesem Werk berechtigt auch ohne besondere Kennzeichnung nicht zu der Annahme, dass solche Namen im Sinne der Warenzeichen- und Markenschutzgesetzgebung als frei zu betrachten wären und daher von jedermann benutzt werden dürften.

Verlag: Südwestdeutscher Verlag für Hochschulschriften GmbH & Co. KG
Heinrich-Böcking-Str. 6-8, 66121 Saarbrücken, Deutschland
Telefon +49 681 37 20 271-1, Telefax +49 681 37 20 271-0
Email: info@svh-verlag.de

Approved by: Zürich, ETH Zürich, Diss., 2011

Herstellung in Deutschland:
Schaltungsdienst Lange o.H.G., Berlin
Books on Demand GmbH, Norderstedt
Reha GmbH, Saarbrücken
Amazon Distribution GmbH, Leipzig
ISBN: 978-3-8381-3023-1

Imprint (only for USA, GB)
Bibliographic information published by the Deutsche Nationalbibliothek: The Deutsche Nationalbibliothek lists this publication in the Deutsche Nationalbibliografie; detailed bibliographic data are available in the Internet at http://dnb.d-nb.de.
Any brand names and product names mentioned in this book are subject to trademark, brand or patent protection and are trademarks or registered trademarks of their respective holders. The use of brand names, product names, common names, trade names, product descriptions etc. even without a particular marking in this works is in no way to be construed to mean that such names may be regarded as unrestricted in respect of trademark and brand protection legislation and could thus be used by anyone.

Publisher: Südwestdeutscher Verlag für Hochschulschriften GmbH & Co. KG
Heinrich-Böcking-Str. 6-8, 66121 Saarbrücken, Germany
Phone +49 681 37 20 271-1, Fax +49 681 37 20 271-0
Email: info@svh-verlag.de

Printed in the U.S.A.
Printed in the U.K. by (see last page)
ISBN: 978-3-8381-3023-1

Copyright © 2011 by the author and Südwestdeutscher Verlag für Hochschulschriften GmbH & Co. KG and licensors
All rights reserved. Saarbrücken 2011

Diss. ETH No. 19778

CHEMICAL WEATHERING AND CALCIUM ISOTOPE FRACTIONATION IN A GLACIATED CATCHMENT

A dissertation submitted to

ETH ZURICH

For the degree of
Doctor of Sciences

Presented by
Ruth Sarah Hindshaw
MChem Edinburgh University
Date of Birth: 10. 3. 1985
Citizen of the
United Kingdom

Accepted on the recommendation of :

Prof. Dr. Bernard Bourdon
Prof. Dr. Ruben Kretzschmar
Dr. Ben Reynolds
Dr. Jan Wiederhold
Prof. Dr. James Kirchner
Prof. Dr. Stephan Kraemer

Zurich, 2011

Contents

List of Figures   VI

List of Tables   VIII

Summary   IX

Zusammenfassung   XI

1 Introduction   **1**
- 1.1 Chemical weathering 1
  - 1.1.1 Small watersheds 3
  - 1.1.2 Soil chronosequences 6
  - 1.1.3 Effect of glaciers on chemical weathering 6
  - 1.1.4 Critical zone observatories 7
- 1.2 Calcium and calcium isotopes 8
  - 1.2.1 Summary of previous work utilising calcium isotopes 10
- 1.3 Research objectives and outline of the thesis 12

2 Field area - Damma Glacier   **15**
- 2.1 Glacier history 15
- 2.2 Geology 16
- 2.3 Soils and vegetation 17
- 2.4 Hydrology and Meteorology 19

3 Hydrological control of stream water chemistry   **23**
- 3.1 Introduction 23
- 3.2 Methods 25
- 3.3 Hydrology and Water Chemistry 25
  - 3.3.1 Catchment meteorology and hydrology 25
  - 3.3.2 Precipitation 26
  - 3.3.3 Correction for precipitation inputs 30
  - 3.3.4 Stream water chemistry 32

	3.4	Discussion: water sources, solute sources and weathering processes	36
		3.4.1 Identifying water sources	36
		3.4.2 Identifying chemical sources of solutes	38
	3.5	Conclusions	47

4 A comparison of weathering fluxes — 49
- 4.1 Introduction — 49
- 4.2 Annual fluxes from the Damma catchment — 51
- 4.3 Comparison of glaciated and non-glaciated samples — 53
  - 4.3.1 Sources of data — 53
  - 4.3.2 The stream water chemical composition of glaciated and non-glaciated catchments — 54
  - 4.3.3 Are silicate weathering fluxes affected by glaciers? — 60
  - 4.3.4 Climate — 61
  - 4.3.5 Lithology — 64
- 4.4 Conclusions — 65

5 Calcium isotopes in a proglacial weathering environment — 67
- 5.1 Introduction — 67
- 5.2 Sample Collection and Results — 69
  - 5.2.1 Rock, Soil and Mineral Separates — 69
  - 5.2.2 Sequential Extractions — 69
  - 5.2.3 Water Samples — 74
  - 5.2.4 Plants — 74
- 5.3 Discussion — 76
  - 5.3.1 Weathering in soils — 76
  - 5.3.2 Porewaters and the exchangeable pool — 77
  - 5.3.3 Stream water — 78
- 5.4 Conclusions — 80

6 Calcium isotope fractionation in alpine plants — 81
- 6.1 Introduction — 81
- 6.2 Sample description — 83
- 6.3 Results — 85
  - 6.3.1 Soil and the soil exchangeable pool — 85
  - 6.3.2 Whole Plant Analyses — 85
  - 6.3.3 Seasonal variation — 87
  - 6.3.4 Variation along the chronosequence — 88
- 6.4 Discussion — 91

		6.4.1	Calcium uptake processes	91
		6.4.2	Within-plant fractionation processes	93
		6.4.3	Intercomparison of alpine plant species	95
		6.4.4	Comparison with other divalent cations	96
		6.4.5	Effect of plants in the biogeochemical Ca cycle	98
	6.5	Conclusions		99

7 Conclusions   **101**

A Methods   **105**

	A.1	Sample Preparation		105
		A.1.1	Soils, rocks and mineral separates	106
		A.1.2	Water samples	107
		A.1.3	Plant samples	107
	A.2	Analytical procedures		108
		A.2.1	Element concentrations	108
		A.2.2	Soil sequential extraction procedure	109
	A.3	Purification of samples for isotope analysis		110
		A.3.1	Calcium	110
		A.3.2	Strontium	112
	A.4	Analysis of Ca isotopes		114
		A.4.1	Radiogenic Ca	116
		A.4.2	Inter-laboratory comparison of Ca isotopic standard measurements	116
	A.5	Analysis of Sr isotopes		118
		A.5.1	MC-ICP-MS	118
		A.5.2	TIMS	118
	A.6	Analysis of O isotopes		119

B Additional figures and datatables   **121**

Bibliography   **133**

Acknowledgements   **153**

List of Figures

1.1	CO_2 and δD from the Vostok ice core	3
1.2	Abundance of the six calcium isotopes	9
2.1	Siegfried map and modern topographical map	16
2.2	Geological map	17
2.3	Map of soil and water sampling locations	18
2.4	Chronosequence development	18
2.5	Overview of catchment meteorology and hydrology during 2008	19
2.6	Photographic time series	20
2.7	Photographs of water sampling locations	21
3.1	Seasonal and diurnal changes in stream water chemistry	26
3.2	Proportions of snow and ice melt and modelled $\delta^{18}O$ over the season	32
3.3	Seasonal and diurnal variations in two element ratios	34
3.4	Seasonal and diurnal variation in $^{87}Sr/^{86}Sr$	35
3.5	Mixing plot of $^{87}Sr/^{86}Sr$ against Ca/Sr	40
3.6	Calculated proportions of surface and sub-glacial meltwater over the season	42
3.7	Conceptual model of glacier hydrology	46
4.1	Conductivity-concentration relationships	52
4.2	Element ratios of granitic and basaltic catchments	59
4.3	Annual cation vs annual Si fluxes	61
4.4	Annual cation fluxes vs runoff and temperature	62
4.5	Arrhenius plot	64
5.1	Ca/Ti and $\delta^{44/42}Ca$ of rocks and soils	71
5.2	Ca isotopic composition of the soils as a function of time	72
5.3	Ternary plot of the sequential extraction data	72
5.4	Ca isotope composition of the different sequential extraction steps	73
5.5	Seasonal Ca flux and isotopic composition of water	75
5.6	Box plot summarising the main pools of calcium analysed.	76

6.1	Ba/Ca and Sr/Ca ratios in leaves	86
6.2	$\delta^{44/42}$Ca variations between plant tissues	87
6.3	Seasonal variation in leaf Ca concentration and $\delta^{44/42}$Ca	89
6.4	Ca concentration and $\delta^{44/42}$Ca of above-ground biomass along the chronosequence	90
6.5	Plot of $\delta^{44/42}$Ca in above-ground biomass against the percentage of roots infected with mycorrhizal fungi	90
6.6	Summary of above-ground Ca isotopic compositions of species analysed	92
6.7	Cartoon of the root tip illustrating proposed mechanisms for Ca isotope fractionation during uptake	94
A.1	Elution curve for a basaltic rock	111
A.2	Elution curve for a synthetic rock standard	113
A.3	Reproducibility of standard measurements	115
B.1	Oxygen isotopic composition of snow	126

List of Tables

2.1	Coordinates of water sampling locations	21
3.1	Major species, ^{87}Sr/^{86}Sr, and δ^{18}O data for seasonal sampling.	27
3.2	Major species, ^{87}Sr/^{86}Sr and δ^{18}O data for diurnal sampling	28
3.3	Major species, ^{87}Sr/^{86}Sr and δ^{18}O data for groundwater (GW) and pore water (PW) samples	29
3.4	Annual element fluxes	30
4.1	Summary of annual fluxes	53
4.2	Summary of flux calculations and precipitation correction methods	55
5.1	Soil data	70
5.2	Analyses of rock, soil and plant samples	71
5.3	Elemental and Ca isotopic compositions of soil sequential extractions	74
5.4	Calcium isotopic composition and concentration data of the water samples	75
6.1	List of plant species analysed in this study	83
6.2	Sr/Ca, Ba/Ca and $\delta^{44/42}$Ca of the soil exchangeable pool	85
6.3	Ca concentrations, $\delta^{44/42}$Ca and element ratios for different plant tissues and soils	88
6.4	$\triangle^{44/42}$Ca$_{plant-soil}$ based on mass balance	88
6.5	Ca concentrations, $\delta^{44/42}$Ca and element ratios for samples collected at various times throughout the season	89
6.6	Ca concentrations, $\delta^{44/42}$Ca and element ratios for plants along the chronosequence	90
6.7	Comparison of stable Ca and Sr isotope ratios relative to soil in *Rhododendron*	97
A.1	Calibration standards for anion analysis	108
A.2	Accuracy and precision of cation (and Si) analysis	109
A.3	Sequential extraction procedure	110
A.4	Full Ca separation procedure	112

A.5	Sr separation procedure	113
A.6	Literature compilation of Ca isotopic standard values	117
B.1	Major species and δ^{18}O data for seasonal sampling (Sites C and D)	122
B.2	Trace metal concentration data for water samples	123
B.3	δ^{18}O values of snow	127
B.4	River data compilation	128

Summary

The chemical weathering of rocks on the Earth's surface is a major geochemical process which not only shapes local landscapes by influencing soil and vegetation types, but is also thought to regulate the Earth's climate over geological timescales. Chemical weathering is affected by many different factors, for example climate, hydrology, tectonic activity and vegetation. Identifying the main processes controlling chemical weathering and predicting chemical weathering rates in different terrains thus requires an interdisciplinary approach combining geology, chemistry, hydrology and biology. The interdisciplinary nature of chemical weathering is reflected in the content of this thesis, which presents work conducted at the Damma glacier forefield in the central Swiss Alps. Soils of different ages are present in the field site as a result of the retreating glacier, allowing both the initial stages of weathering and accompanying soil formation to be studied. This thesis was conducted as part of a larger inter-disciplinary project at this field site (BigLink) which investigated initial weathering, soil formation and ecosystem development.

Chemical weathering strongly influences stream water chemistry and temporal variations in stream water chemistry can be used to identify the environmental factors controlling weathering processes. The causes of temporal variability in stream water chemistry at the Damma glacier catchment are discussed in chapter 3. A hydrological model was used to calculate the relative contributions of the principal water sources, snow melt and ice melt, throughout the period of this study. Significant systematic seasonal and diurnal variations were observed in the stream water chemistry, which cannot be caused by the mixing of water draining different lithologies. Pronounced seasonal minima in $\delta^{18}O$ and $^{87}Sr/^{86}Sr$ were attributed to spring snow melt. Clear changes in major cation to Si ratios between summer and winter were observed, with high ratios in summer and low ratios in winter. These changes were interpreted to reflect seasonal changes in the average residence time of water in the sub-glacial drainage network with short residence times in summer, when the discharge was greatest, and long residence times in winter, when the discharge was lowest. This thesis shows that the time dependent stoichiometry of cation to Si ratios in glacial stream water (and likely all catchments) strongly depends on the hydrological state of the catchment at the time of sampling. Annual fluxes based on spot samples varied by a factor of six depending on the time of year in which the sample was collected, highlighting the importance of long-term catchment monitoring in order to precisely quantify silicate weathering processes. A compilation of annual fluxes from granitic catchments (chapter 4) highlighted the fact that glaciers do not enhance weathering rates compared to non-glaciated catchments with similar discharge. However, the chemical composition of stream water is altered in response to the physical erosion regime.

The weathering of calcium (Ca) from silicate minerals is strongly coupled to the carbon cycle through the consumption of carbon dioxide during chemical weathering and the formation of calcium carbonate in the ocean. Stable calcium isotopes have the potential to provide additional constraints on the Ca biogeochemical cycle by tracing the influence of processes which induce Ca isotope fractionation. Although there has been considerable interest in the use of calcium isotopes as a paleo-oceanographic tracer of past climate there has been relatively little research focussing on modern day terrestrial ecosystems, despite the influence of weathering processes on the oceanic Ca isotopic composition. This thesis presents a detailed study of calcium isotope fractionation along a soil chronosequence (chapter 5). Intensive sampling of rocks, stream water and soils of different ages revealed negligible Ca isotope fractionation. This lack of variation indicates that the dissolution of the bulk silicate rock does not strongly fractionate Ca isotopes. The only Ca pool which was strongly fractionated relative to bulk rock was vegetation, which exhibited an enrichment of light Ca isotopes. For this reason, Ca isotope fractionation in several species of alpine plants from this field site was investigated in more detail (chapter 6). It was found that there was significant Ca isotope fractionation between soil and root tissue and the magnitude of this fractionation was similar in all investigated plant species. The large Ca isotopic difference between root tissue and soil is probably caused by the preferential binding of light Ca isotopes to root adsorption sites. Within-plant Ca isotope fractionation, however, was species dependent. For example, in samples of leaf tissue collected throughout the growing season, Ca isotope ratios increased with leaf age in woody species but remained constant in herbs and grasses. Whole plant Ca isotopic compositions were largely determined by the root cation exchange capacity, but species-specific factors such as the presence of a woody stem or Ca oxalate are also likely to contribute to the observed differences in whole plant Ca isotopic compositions both between and within species. The use of calcium isotopes to trace biogeochemical processes has most potential in watersheds where the Ca biogeochemical cycle is dominated by secondary processes such as biological cycling, adsorption and secondary mineral precipitation, and not by mineral weathering, which does not fractionate Ca isotopes. The impact of biological cycling will depend on the species present and the stage of vegetation succession.

The results presented in this thesis stress the importance of a 'whole-system' approach to the study of chemical weathering, especially if the myriad natural variables affecting this complex process are to be successfully disentangled and quantified.

Zusammenfassung

Chemische Verwitterung von Gestein an der Erdoberfläche ist ein wichtiger geochemischer Prozess, welcher, durch seinen Einfluss auf den Boden und die Vegetation, nicht nur Landschaften formt, sondern über geologische Zeiträume auch das Klima reguliert. Chemische Verwitterung wird von verschiedenen Parametern, wie klimatischen und hydrologischen Faktoren, tektonischer Aktivität und Vegetation beeinflusst. Um die entscheidenden Prozesse der chemischen Verwitterung zu identifizieren ist daher ein interdisziplinärer Ansatz aus Geologie, Chemie, Hydrologie und Biologie notwendig. Dieser interdisziplinäre Ansatz wird durch den Aufbau dieser Doktorarbeit widergespiegelt. Die vorliegende Arbeit präsentiert Ergebnisse, die von Proben aus dem Dammagletschervorfeld in den Schweizer Zentralalpen gewonnen wurden. Durch den sich zurückziehenden Gletscher sind an diesem Gletschervorfeld Böden verschiedener Altersstufen vorhanden. Dies ermöglicht das Studium der frühen Phasen der chemischen Verwitterung und der damit einhergehenden Bodenzusammensetzung. Diese Doktorarbeit wurde innerhalb eines grösseren, interdisziplinären Projektes (BigLink) am Dammagletschervorfeld durchgeführt, welches die frühen Phasen der Verwitterung, Bodenbildung und Entwicklung des lokalen Ökosystems untersucht.

Verwitterungsprozesse haben einen grossen Einfluss auf die chemische Zusammensetzung von Flusswasser und die zeitlichen Änderungen dieser Zusammensetzung geben Aufschlüsse über die exogenen Faktoren, welche die Verwitterung beeinflussen. Die Ursachen der zeitlichen Schwankungen in der chemischen Zusammensetzung des Wassers aus dem Damma-Einzugsgebiet werden in Kapitel 3 diskutiert. Um die relativen Beiträge der wesentlichen Wasserquellen, Schmelzwasser von Schnee und Gletschereis, über den Beobachtungszeitraum zu berechnen, wurde ein hydrologisches Model verwendet. Signifikante, systematische Schwankungen über den Jahres- und Tagesverlauf wurden beobachtet, welche nicht auf die Durchmischung von Wasser aus verschiedenen Gesteinsarten zurückgeführt werden können. Deutliche Jahrestiefstwerte im $\delta^{18}O$ Gehalt und im Verhältnis der Strontiumisotope $^{87}Sr/^{86}Sr$ im Frühjahr wurden der Schneeschmelze zugeordnet. Desweiteren wurden klare Änderungen im Verhältnis der Hauptkationen zu Silizium mit hohen Werten im Sommer und niedrigen Werten im Winter beobachtet. Diese Änderungen wurden dahingehend interpretiert, dass sie die jahreszeitlichen Schwankungen in der Aufenthaltszeit des Wassers im subglazialen

Entwässerungsnetzwerk widerspiegeln, wobei die Aufenthaltszeit im Sommer, bei hohem Wasserabfluss, am kürzesten, und im Winter, bei schwachem Abfluss, am längsten war. Diese Doktorarbeit belegt, dass die zeitliche Änderung im Verhältnis von Kationen zu Silizium in vergletscherten, und wahrscheinlich allen, Wassereinzugsgebieten stark vom hydrologischen Zustand des Einzugsgebiets zum Zeitpunkt der Probennahme abhängt. Die jährlichen Kationenflussraten, berechnet auf Basis der einzelnen Proben, variierten um einen Faktor sechs, je nachdem wann im Jahresverlauf die Probe genommen wurde. Dies unterstreicht die Wichtigkeit einer längeren Beobachtung des Wassereinzugsgebietes, soll der Silikatverwitterungsprozess präzise quantifiziert werden. Eine Zusammenstellung der jährlichen Flussraten der granithaltigen Einzugsgebiete (Kapitel 4) zeigt, dass der Gletscher, verglichen mit nicht vergletscherten Einzugsgebieten, nicht zu erhöhten Zersetzungsraten führt. Allerdings wird die chemische Zusammensetzung des Flusswassers durch die physikalische Erosion verändert.

Die Verwitterung von Calcium (Ca) aus Silikaten ist durch den Verbrauch von Kohlenstoffdioxid während der chemischen Verwitterung und die Bildung von Calciumcarbonaten in den Ozeanen, stark an den Kohlenstoffzyklus gebunden. Stabile Calciumisotope ermöglichen weitere Einsichten in den biogeochemischen Calciumkreislauf, da sie es erlauben Prozesse zu verfolgen welche zu Fraktionierung von Calciumisotopen führen. Obwohl Calciumisotope häufig als paleo-ozeanographische Klimaindikatoren Verwendung finden, gibt es, trotz der Bedeutung der Verwitterungsprozesse auf das ozeanische Ca-Isotopenverhältnis, nur vergleichsweise wenige Forschungsansätze für heutige terrestrische Ökosysteme. Diese Doktorarbeit stellt die erste detaillierte Studie über Ca-Isotopenfraktionierung im Zusammenhang mit einer Chronosequenz des Bodens dar (Kapitel 5). Umfangreiche Gesteins-, Flusswasser- und Bodenproben verschiedenen Alters zeigen vernachlässigbare Ca-Isotopenfraktionierung. Die fehlende Varianz im Ca-Isotopenverhältnis deutet darauf hin, dass die Lösung von silicathaltigem Gestein zu keiner starken Fraktionierung der Ca-Isotope führt. Nur in den Pflanzenproben, welche Anreicherungen von leichten Ca-Isotopen aufweisen, wurde eine starke Fraktionierung festgestellt. Aus diesem Grund wurde die Ca-Isotopenfraktionierung in verschiedenen Arten von alpinen Pflanzen genauer untersucht (Kapitel 6). Es zeigte sich ein deutlicher Unterschied im Isotopenverhältnis zwischen den Bodenproben und dem Wurzelgewebe. Die Stärke dieser Fraktionierung war vergleichbar in allen untersuchten Pflanzenarten. Die grosse Differenz im Isotopenverhältnis zwischen Boden und Wurzelgewebe wird wahrscheinlich durch die bevorzugte Absorption von leichten Ca-Isotopen durch das Wurzelgewebe hervorgerufen. Die Isotopenverteilung innerhalb der Pflanzen war jedoch artabhängig. Zum Beispiel stieg das Ca-Isotopenverhältnis in den Blattgewebeproben, welche über die gesamte Wachstumsphase genommen wurden, mit zunehmendem Alter in Baumblättern an, blieb jedoch für Kräuter und Gräser gleich. Das Isotopenverhältnis innerhalb der gesamten Pflanze wurde hauptsächlich durch die

Kationenaustauschkapazität bestimmt. Wahrscheinlich tragen allerdings auch artspezifische Faktoren, wie das Vorhandensein eines hölzernen Stammes oder die Einlagerung von Calciumoxalat, zu den gefundenen Unterschieden im Ca-Isotopenverhältnis zwischen den verschiedenen Pflanzenarten wie auch innerhalb einer Art bei. Die Verwendung von Calciumisotopen zur Untersuchung von biogeochemischen Prozessen hat ihr grösstes Potential in Wassereinzugsgebieten in denen der biogeochemische Calciumkreislauf durch sekundäre Prozesse wie biologische Kreisläufe, Adsorption und sekundäre Mineralablagerungen bestimmt wird, und nicht durch Mineralverwitterung, welche nicht zur Ca-Isotopenfraktionierung führt. Der Einfluss auf die biologischen Kreisläufe ist abhängig von den vorhandenen Arten und dem Zustand der Vegetationsfolge.

Die präsentierten Resultate unterstreichen die Wichtigkeit eines ganzheitlichen Ansatzes bei der Untersuchung der chemischen Verwitterung, insbesondere wenn die vielen natürlichen Parameter, welche diesen Prozess beeinflussen, erfolgreich bestimmt und quantifiziert werden sollen.

Chapter 1

Introduction

1.1 Chemical weathering

The chemical weathering of rocks forms one part of the great geological cycle where rocks are continually being formed and broken down again, shaping the Earth's landscape. Chemical weathering releases elements which were previously contained in the various constituent minerals of rocks, providing nutrients for living organisms, the raw materials for soil formation and defining the chemical composition of rivers. The biogeochemical cycle of nearly every element is affected by weathering. The primary weathering agents are protons, which are commonly derived from the dissolution of carbon dioxide gas in water (Eqn. 1.1), but can also be derived through other chemical reactions such as the oxidation of sulphide minerals and organic matter.

$$CO_{2(g)} + H_2O_{(l)} \rightleftharpoons H_2CO_{3(aq)} \rightleftharpoons HCO_{3(aq)}^- + H_{(aq)}^+ \tag{1.1}$$

Acid hydrolysis of minerals occurs when they react with protons. The corresponding reactions for a silicate mineral such as anorthite and a carbonate mineral such as calcite can be written as follows:

$$CaAl_2Si_2O_{8(s)} + 2CO_{2(g)} + 3H_2O_{(l)} \rightleftharpoons Ca_{(aq)}^{2+} + 2HCO_{3(aq)}^- + Al_2Si_2O_5(OH)_{4(s)} \tag{1.2}$$

$$CaCO_{3(s)} + H_2O_{(l)} + CO_{2(g)} \rightleftharpoons Ca_{(aq)}^{2+} + 2HCO_{3(aq)}^- \tag{1.3}$$

Unlike the reaction for carbonates (Eqn. 1.3), the weathering of silicates (Eqn. 1.2) rarely results in the total dissolution of the initial mineral. Instead, secondary minerals form (kaolinite in Eqn. 1.2); the exact composition being dependent on specific environmental conditions such as pH. Regardless of the type of silicate mineral, weathering results in the release of cations and bicarbonate into the environment, coupled to the consumption of atmospheric CO_2. Once the cations and bicarbonate released during weathering reach the ocean, the reverse reaction of equation 1.3 occurs, forming calcite or its polymorph aragonite when the reacting cation is Ca and magnesite ($MgCO_3$) when the reacting cation is Mg.

Carbon dioxide is a greenhouse gas and data from ice cores have revealed synchronous cyclical trends in the levels of atmospheric CO_2 and temperature over the last 420 kyr [Petit et al., 1999], suggesting a strong coupling between these two parameters (Fig. 1.1). Further in the past, climate is inferred from proxies such as Mg/Ca ratios, $\delta^{18}O$ and $\delta^{13}C$ obtained from foraminifera retrieved from deep sea sediment cores [Lear et al., 2000, Zachos et al., 2001]. The cyclicity observed in Earth's climate is thought to be initiated by changes in solar insolation due to changes in the Earth's orbit, known as Milankovitch cycles [Imbrie et al., 1992]. Milankovitch cycles alone cannot explain past climate changes: in addition, a complex series of feedback mechanisms controls the magnitude and duration of each cycle [Imbrie et al., 1993]. Changes in climate on a short timescale (e.g. Quaternary glacial-interglacial cycles) are thought to be due to feedbacks occurring in the oceans and the biosphere, affecting the distribution of carbon between these two reservoirs and the atmosphere [Brook et al., 1996, Sigman et al., 2010]. Over longer timescales (> 1000 kyr) different feedback mechanisms operate which result in more gradual changes to the carbon cycle [Sundquist and Visser, 2003]. One potential feedback is between weathering and climate [Walker et al., 1981, Berner et al., 1983]. The Earth's crust is by far the largest reservoir of carbon, but fluxes of carbon from this reservoir, via the weathering of rocks, are more than an order of magnitude less than those between the ocean and the atmosphere. Thus, changes in weathering rates are only manifested over long periods of time. The type of rock which is weathered must be considered: the weathering of carbonates can only affect climate on short timescales since atmospheric CO_2 is only removed for the length of time it takes the solutes to reach the ocean. In contrast, silicate weathering results in the net removal of one mole of CO_2 for each mole of mineral weathered. The time of removal from the atmosphere is the length of time for the complete subduction and exhumation cycle.

As a result of the potential for long-term buffering of global climate by silicate weathering, there has been considerable effort to quantify the amount of silicate weathering occurring in the present day and the factors affecting weathering rates, in order to model past CO_2 changes [Ludwig et al., 1999, Berner and Kothavala, 2001]. There are three main approaches to the investigation of silicate weathering (and weathering in general): river studies, soil profiles and laboratory studies. Rivers transfer weathering products from the continents to the oceans and are an 'instantaneous' measure of weathering. By analysing the chemical composition of a river it is possible to calculate a chemical weathering flux for the whole catchment, and by measuring the largest rivers in the world, global weathering fluxes can be estimated. The proportion of the total weathering flux contributed by silicate weathering can be estimated using element ratios and isotope

Chemical weathering 3

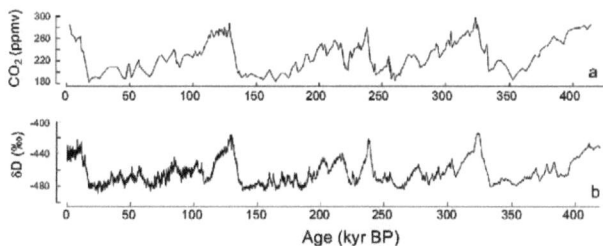

Figure 1.1: CO_2 (a) and δD (b) from the Vostok ice core [Petit et al., 1999]. δD is a proxy for local temperature. Note the synchronous behaviour of these two parameters over time. Low CO_2 and δD values correspond to glacial periods e.g. 150 kyr BP.

tracers [e.g. Gaillardet et al., 1999]. Soil profiles record a vertical profile of weathering from unweathered bedrock at the base to the overlying mature soil at the top and provide weathering rates integrated over time [e.g. Amundson, 2003]. The thickness of the soil profile is a function of the weathering rate and the surface erosion rate and by measuring the depletion of mobile elements relative to immobile elements with depth, chemical weathering rates can be calculated. Laboratory studies enable the calculation of specific mineral and rock weathering rates under controlled conditions [e.g. Brantley, 2003]. The calculated mineral weathering rates can then be compared to field derived weathering rates and models of weathering [Lasaga et al., 1994, White and Brantley, 2003]. Obtaining weathering rates from stream water chemistry in small watersheds and the use of soil chronosequences to obtain soil weathering rates are discussed in more detail in the following two sections.

1.1.1 Small watersheds

The stream water chemistry of numerous small (<50 km^2) watersheds has been reported in the scientific literature. Small watershed studies are extremely valuable because local conditions can easily be monitored and detailed characterisation of local geology and vegetation is often possible, allowing biogeochemical processes to be investigated in detail. Early studies were aimed at quantifying the annual inputs and outputs to a particular watershed in order to assess nutrient budgets and quantify element cycling within the watershed. A positive net output flux could be attributed to weathering reactions and a negative net output flux was attributed to biological uptake. Such studies highlighted the detrimental effects of acid rain deposition in many forested catchments of northern Europe and America, for example, the accelerated loss of base cation nutrients such as Ca [Pačes, 1985, April et al., 1986]. More detailed models of element fluxes were able to be developed for more intensively studied catchments, e.g. the model of Ca fluxes for Hubbard Brook Experimental Forest (HBEF, USA) includes terms for each soil horizon and dead and living biomass [Likens et al., 1998].

The weathering component of small watershed mass balance models can be broken down into individual mineral components. Here, the aim is to identify exactly which minerals are weathering. The dominant approach is the chemical mass balance method of Garrels and Mackenzie [1967]. This approach assumes that the difference in chemical composition between rain water (input) and stream water (output) is the result of chemical weathering reactions. Inverse modelling determines how much of each mineral must have reacted to result in the observed stream water composition using a set of simultaneous equations. Annual element fluxes are preferred as input parameters over individual concentration measurements since over this period the uptake and release of elements by ion exchange and vegetation is assumed to be in balance and can therefore be neglected [Finley and Drever, 1997]. There are often several possible model outputs because there are many ways to produce the same output solution by changing the relative composition of minerals reacting. Identifying a realistic model output depends on a detailed knowledge of the potential input mineral phases and the saturation states of minerals in the system. Isotope tracers can help identify which minerals are weathering, providing each mineral has a distinct isotopic ratio. Strontium isotopes are the most frequently used isotope tracer [Blum et al., 1994, Bullen et al., 1996, Clow et al., 1997]. Nevertheless, inverse modelling assumes that minerals weather congruently and does not take into account kinetic factors which affect mineral weathering such as hydrology. Thus, although inverse modelling can provide an initial overview of potential mineral reactions it does not necessarily represent the mineral reactions which are actually occurring [Bullen et al., 1996]. It has long been noted that inverse modelling results in an apparent excess Ca in granitic catchments. One solution has been to attribute this excess to the importance of the weathering of trace carbonate phases [Mast et al., 1990]. Other plausible explanations include the weathering of calcic silicates such as epidote [Oliva et al., 2004] or apatite [Aubert et al., 2002] and the incongruent weathering of plagioclase. Identifying whether the source of Ca is silicate or carbonate is vital for determining the ratio of carbonate to silicate weathering in watersheds.

Small watersheds have also been used extensively to understand catchment hydrochemistry: the identification of water sources and how these change over time. A common technique is hydrograph separation based on end-member mixing, where potential end-member water compositions are identified e.g. groundwater and rain, and the relative contributions of each source to the stream output are calculated [Liu et al., 2004, Brown et al., 2006]. The main drawback of this approach is that it assumes spatially and temporally fixed end-member compositions, which is not always valid e.g. the chemical composition of snow changes during snowmelt, though statistical methods have been developed to take into account some of the uncertainty in source compositions [Soulsby et al., 2003]. Water isotopes (δD, $\delta^{18}O$) give information on the original water sources but do not give information on the flowpaths of the water through the catchment. To identify flowpaths, additional chemical tracers are required. Identifying flowpaths is important for solving long-standing hydrological problems e.g. how rain water alters stream chemical compositions [Kirchner, 2003]. Ideally these tracers should be conservative (e.g. $^{87}Sr/^{86}Sr$), but non-conservative tracers such as element ratios are also useful [Hogan and Blum, 2003]. Multi-tracer approaches offer the most promising approach to understanding the relationship between hydrology and stream water chemistry.

A paired catchment approach is often used in small watershed studies. Two nearly identical catchments are chosen and one is 'disturbed'. Alternatively, one catchment is studied and baseline data is collected for several years prior to the 'disturbance'. The resultant changes in stream water chemistry (and other catchment properties) are then monitored. Three examples of this approach are given below:

- In 1999, the mineral wollastonite ($CaSiO_3$) was added to a sub-catchment of the HBEF. Wollastonite had distinct element and Sr isotope ratios to naturally occurring sources in the catchment allowing uptake rates of Ca by different tree species to be calculated [Dasch et al., 2006] and highlighted the seasonal variability in the contribution of different hydrological flowpaths to the overall stream water chemistry [Nezat et al., 2010].

- The importance of vegetation in controlling discharge, stream water chemistry and total element fluxes from catchments has been shown in studies where vegetation has been completely removed from the catchment [Likens et al., 1970, Balogh-Brunstad et al., 2008].

- Artificial rain experiments using sprinkler systems have allowed the investigation of the mechanisms by which increased discharge increases weathering rates [Clow and Drever, 1996] and the addition of chemical tracers (Br^-, ^2H) to the sprinkler water has enabled the movement of the added water through the catchment to be traced [Anderson et al., 1997b].

Compilations of annual fluxes from many small catchments which have the same lithology allows the assessment of parameters such as precipitation and temperature on weathering rates. One of the first compilations considered 101 different rivers draining a variety of rock types [Bluth and Kump, 1994]. This study clearly showed that lithology is a primary control of weathering rates followed by runoff, though at high runoff values the influence of lithology was diminished. Focussing on granite, a compilation of 68 granitic watersheds by White and Blum [1995] found that weathering rates were dependent on the amount of precipitation and mean annual air temperature. An expanded study of 99 granitic watersheds [Oliva et al., 2003] reached the similar conclusion that weathering rates were dependent on temperature and runoff. A dependence of temperature and runoff on weathering rates was also observed for basaltic catchments [Dessert et al., 2001]. In contrast, Gaillardet et al. [1999] concluded that temperature and runoff could not be used to predict silicate weathering rates. Similarly, West et al. [2005] showed that climatic parameters (temperature and runoff) only affected weathering rates in weathering limited (material is removed before weathering reactions are complete) settings based on a compilation of 18 regions. In other (transport limited) catchments, physical erosion was found to be the dominant factor controlling weathering rates. The importance of physical erosion was also identified by Millot et al. [2002] and Riebe et al. [2004]. Developing models which can predict weathering rates is required to model the response of weathering to climatic and tectonic changes during Earth's history.

1.1.2 Soil chronosequences

A soil chronosequence provides information on the rate of soil formation (and therefore weathering rates) and the change of soil properties with time. An ideal soil chronosequence is a series of soils where vegetation, climate, topography and bedrock have remained constant and the only difference between the soils is the age [Jenny, 1941]. In reality, factors like vegetation and climate will likely also have varied throughout the time of soil development. The majority of soil chronosequences can be classified as 'post-incisive' - something produces a series of progressively younger surfaces [Huggett, 1998], examples are an incising river producing alluvial terraces and a retreating glacier exposing new ground.

Several soil chronosequence studies have shown that weathering rates are highest in the youngest soils and decrease exponentially to a steady state after several thousand years [Bain et al., 1993, Taylor and Blum, 1995, White et al., 1996, Egli et al., 2001]. In catchments containing both carbonate and silicate rocks, the ratio of carbonate to silicate weathering was found to be high in young soils and only decreased after several thousand years [Jacobsen et al., 2002]. However, Anderson et al. [2000] found that silicate weathering could become important as soon as vegetation was established, i.e. within a few hundred years. Soil chronosequences encompassing the first few hundred years of soil formation demonstrate measurable losses of mineral phases such as epidote and biotite [Mavris et al., 2010]. Although one study found biotite weathering enhanced in young soils compared to older ones [Blum and Erel, 1997], similar changes in soil chemistry could have been caused by potassium feldspar weathering [Bullen et al., 1997]. These apparently conflicting studies highlight the problem of unambiguously identifying which minerals are weathering.

1.1.3 Effect of glaciers on chemical weathering

The Earth's climate has oscillated between periods with extensive glacial cover and periods with minimal glacial cover (Fig. 1.1). In order to understand whether increased

Chemical weathering

glacial cover affects weathering rates (and thus atmospheric CO_2 levels, Eqn. 1.2), it is important to compare modern glaciated catchments with unglaciated ones.

There are three main types of glacier based on the thermal regime at the base of the glacier: cold, polythermal and warm [Tranter, 2003]. In a cold-based glacier, there is negligible liquid water at the rock-ice interface and weathering is limited to the ice margins. When basal ice exceeds the pressure melting point, liquid water exists at the bed and the glacier is classified as warm-based. Warm-based glaciers develop complex sub-glacial drainage structures which change throughout the melt season [Nienow et al., 1996]. Weathering can occur throughout the sub-glacial channel network. At higher latitudes, glaciers can be polythermal, having a warm-based interior and cold-based margins.

Early studies of the weathering fluxes from temperate (warm-based) glacial catchments suggested that glaciers enhanced weathering rates due to high discharge, high physical erosion rates and highly reactive glacial flour which offset any reduction in weathering as a result of low temperatures [Reynolds and Johnson, 1972, Metcalf, 1986]. This view has been supported by more recent studies [Sharp et al., 1995, Oliva et al., 2003], although several other studies have concluded that, compared to non-glaciated catchments, silicate weathering rates are reduced or are unaffected in glaciated catchments [White and Blum, 1995, Gíslason et al., 1996, Anderson et al., 2000, Hosein et al., 2004]. Studies of cold-based glaciers indicate that weathering rates are low compared to comparable polythermal or warm-based glaciers [Hodgkins et al., 1997, 1998, Hodson et al., 2000], but this may simply be a function of discharge or lithology [Hodson et al., 2000]. The solute chemistry of cold-based glaciers is controlled by snow melt, weathering in lateral moraines and icing (frozen meltwater, concentrated in solutes). The weathering rates of polythermal glaciers are less than, or equal to, those of warm-based glaciers [Wadham et al., 1997]. It is possible that in the early melt-season, the cold-based edge blocks meltwater transport from the interior allowing increased weathering to occur [Tranter, 2003].

Basal regimes can change if the glacier changes in size, for example, if a polythermal glacier becomes too thin it will change to a cold-based regime. Present weathering rates may be affected by the nature of the glacier in the past [Hodgkins et al., 1998]. It has also been suggested that weathering rates in deglaciated areas continue to be influenced by the landscape changes caused by the glacier (fractured bed-rock, moraine deposits, glacial sediment), long after the glacier itself has disappeared [Vance et al., 2009].

1.1.4 Critical zone observatories

The critical zone is defined as the external terrestrial layer extending from the outer limits of vegetation down to and including the zone of groundwater, including all the biological, chemical, hydrological and geological processes occurring inbetween [Brantley et al.,

2005]. The study of the 'critical zone' is of course not new but the effort to link several different disciplines to gain a more holistic view of weathering is. For example, a geochemist may interpret a change in stream water chemistry as a change in mineral weathering reactions whereas a hydrologist may interpret it as a change in a water source. By considering stream water chemistry as a mixture of water sources *and* as the product of different minerals weathering, a more detailed understanding of catchment hydrogeochemical processes should result.

Weathering is an inherently complex process, with several factors operating at once. By establishing a network of carefully selected sites and building on the knowledge already gained from previous small watershed studies, it is hoped that the influence of individual variables such as lithology and precipitation can be ascertained, in much the same way as a chronosequence allows the investigation of temporal changes. To date, most sites are located in the US, and this thesis concerns one of the 'international' sites located in the Swiss Alps (chapter 2). This site represents granitic catchments in the early stages of soil development.

1.2 Calcium and calcium isotopes

Calcium, an alkaline earth metal, is the fifth most abundant element in the Earth's crust. Calcium forms a key constituent of many main rock forming minerals such as calcite and plagioclase. In addition, it is an essential nutrient for all living organisms, except perhaps fungi [Marschner, 1995]. Many living organisms depend on Ca for structural purposes such as forming bones and teeth ($Ca_5(PO_4)_3(OH)$, hydroxylapatite) and shells ($CaCO_3$, calcium carbonate). Furthermore, the geochemical cycles of calcium and carbon are intimately linked (for example, through the formation of $CaCO_3$ in the ocean). The analysis of calcium isotopes provides a new tool to trace and quantify fluxes within the calcium biogeochemical cycle.

Calcium has six naturally occurring stable isotopes, of which ^{40}Ca is the most abundant (Fig. 1.2). Absolute stable isotope ratios vary in nature since the different isotopes react differently. For example, in kinetic reactions the light isotopes will be preferentially incorporated into the reaction products since this is most energetically favourable. Thus, differences in isotopic composition between a starting material and an end material can be used to infer which processes acted on that element. This type of information is not obtainable from concentration data. Calcium has one radiogenic isotope: ^{40}Ca. ^{40}K decays to ^{40}Ar via electron capture and ^{40}Ca via beta decay, 89.52% of ^{40}K atoms decay to ^{40}Ca with a half-life of 1.248×10^9 yr. The long half-life and naturally high abundance of ^{40}Ca has prevented the widespread use of the K-Ca dating system as a geochronometer and limited its use to very old rocks with high K/Ca ratios.

Calcium isotopes belong to a family of 'nontraditional' stable isotope systems (the

Calcium and calcium isotopes

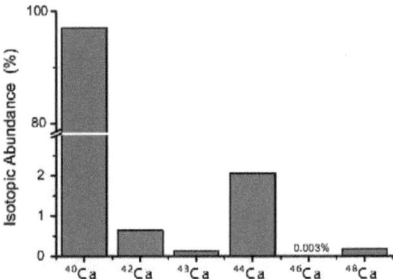

Figure 1.2: Abundances of the six natural calcium isotopes, note the scale break. At 96.98%, ^{40}Ca is overwhelmingly the most abundant isotope.

'traditional' systems being H, C, N, O and S). These are isotope systems in which isotopic differences in natural materials have only relatively recently been able to be resolved due to the vast improvements in measurement techniques. Nontraditional stable isotopes encompass many elements such as Ca, Mg, Fe and Si which are major elements in earth system cycles. Current research aims to use the fractionation observed in these isotope systems to obtain a greater understanding of low temperature biogeochemical processes occurring at the Earth's surface.

Stable calcium isotope variations are expressed as the change in the ratio of two isotopes relative to a standard. These variations are small and are reported in delta notation i.e. the ratio multiplied by 1000, with units of per mil (‰) (equation 1.4, x=0 or 2).

$$\delta^{44/4x}Ca = 1000 \left\{ \frac{\left(\frac{^{44}Ca}{^{4x}Ca}\right)_{sample}}{\left(\frac{^{44}Ca}{^{4x}Ca}\right)_{standard}} - 1 \right\} \quad (1.4)$$

Both the $\delta^{44/42}$Ca and the $\delta^{44/40}$Ca ratios are reported in the literature. Although the magnitude of fractionation in $\delta^{44/40}$Ca ratios is approximately double that of $\delta^{44/42}$Ca ratios (due to larger relative mass difference), $\delta^{44/42}$Ca is not affected by radiogenic anomalies from ^{40}Ca. Studies which use $\delta^{44/40}$Ca must assume that radiogenic anomalies are negligible. Three main standards are in current use: seawater, NIST SRM 915a and 'bulk silicate earth', despite attempts to agree on one main standard [Eisenhauer et al., 2004]. All stable isotope variations reported in this thesis are relative to NIST SRM 915a. Radiogenic ^{40}Ca anomalies are reported in epsilon units: the relative difference of the mass-fractionation corrected ratio and a defined value ($^{40/42}$Ca = 151.029) in parts per ten thousand (equation 1.5). The $^{42/44}$Ca normalization ratio is taken to be 0.31221 after

DePaolo [2004].

$$\epsilon_{Ca} = 10000 \left\{ \frac{(\frac{40Ca}{42Ca})_{sample}}{151.029} - 1 \right\} \quad (1.5)$$

1.2.1 Summary of previous work utilising calcium isotopes

The first measurements of Ca isotope ratios were made in the 1950s [a summary of early work can be found in Corless, 1968]. Although early work hinted at measurable Ca isotope variations, reproducibility was poor and the errors were large, in part due to the technical problems of measuring the minor Ca isotopes (Fig. 1.2). Calcium isotopes are measured by mass spectrometry - isotopes are separated according to mass and 'counted' in collectors. The mass spectrometry procedure itself causes isotope fractionation such that the measured isotopic composition is not identical to the original sample. The development of double-spike techniques allowed instrumental mass fractionation to be accounted for and enabled more accurate and precise measurements of Ca isotope ratios [Russell et al., 1978b]. The double-spike technique involves adding a spike to the sample which is enriched in two isotopes. The composition of the spike is known and using an iterative procedure the sample isotopic composition can be extracted from that of the measured mixture. A double-spike also permits the correction of any isotope fractionation which occurs during pre-measurement chemical separation procedures. Nowadays, the double-spike technique is used predominantly for thermal ionisation mass spectrometry (TIMS) measurements. Calcium isotopes are also measured by multi-collector inductively coupled plasma mass spectrometry (MC-ICP-MS). With this technique the most abundant isotope, ^{40}Ca, cannot be measured due to isobaric interference from ^{40}Ar. Instrumental mass fractionation in MC-ICP-MS is corrected for by standard-sample bracketing: the drift of the standard with respect to a pre-defined delta value ($\delta^{44/42}$Ca) enables the correction of sample measurements. Further discussion of measurement techniques can be found in Fantle and Bullen [2009], Boulyga [2010] and appendix A.4.

Calcium isotopes have found use in a wide range of fields including cosmochemistry [^{48}Ca nucleosynthetic anomalies, Jungck et al., 1984], geochronology [radiogenic ^{40}Ca, Marshall and DePaolo, 1982, 1989, Nelson and McCulloch, 1989, Fletcher et al., 1997, Kreissig and Elliott, 2005, Harrison et al., 2010] and stable isotope fractionation. The majority of stable Ca isotope literature focusses on the stable Ca isotopic compositions of calcium carbonate bearing organisms in the modern ocean and in ocean sediments. The calcium isotopic composition of the modern ocean can be assumed to be homogeneous and reflects the long-term balance between riverine and hydrothermal inputs and removal through carbonate precipitation. It is thought that the Ca isotopic composition of marine carbonates reflects that of seawater, thus analysis of carbonates in marine sediment cores can be used to create a record of the oceanic Ca isotopic composition through time. Variations in the Ca isotope ratio of seawater provide information on the balance between sources (e.g. weathering) and sinks (e.g. sedimentation) of oceanic Ca in the past, reflecting changes in climate and tectonic events [Zhu and Macdougall, 1998, De La Rocha and DePaolo, 2000, Schmitt et al., 2003a, Fantle and DePaolo, 2005, Farkaš et al., 2007]. Understanding the environmental controls (e.g. temperature and species) on the Ca isotopic difference between seawater and calcifying organisms and the differences between organic and inorganic carbonate precipitation form a related focus of current research using Ca isotopes [Nägler et al., 2000, Gussone et al., 2003, Chang et al., 2004, Lemarchand et al., 2004, Sime et al., 2005, Jacobson and Holmden, 2008, Holmden, 2009].

Measuring Ca isotopes in terrestrial organisms is a growing area of research with potential for tracing trophic level [Skulan et al., 1997], changes in bone mass and sources of dietary calcium [Skulan and DePaolo, 1999, Chu et al., 2006, Skulan et al., 2007, Heuser and Eisenhauer, 2010, Reynard et al., 2010], with potential archaeological applications [Reynard et al., 2010, Heuser et al., 2011]. Further applications of Ca isotopes include: quantifying Ca isotope fractionation during diffusion in water [Bourg et al., 2010] and silicate melts [Watkins et al., 2009]; investigating calcium isotope fractionation during igneous (high temperature) processes [Amini et al., 2009, Huang et al., 2010]; investigating planetary formation by comparing stable and radiogenic Ca isotopes in meteorites with terrestrial rocks [Simon et al., 2009, Simon and DePaolo, 2010]; further constraining the oceanic Ca cycle using radiogenic Ca [Caro et al., 2010] and Ca isotope fractionation in hydrothermal fluids [Amini et al., 2008].

Despite the need to quantify the Ca isotopic composition of continental inputs to the ocean, and the potential for using Ca isotopes to understand biogeochemical Ca cycling, there have been relatively few studies of Ca isotopes in terrestrial materials such as soils, rivers and plants. Initial data from the Strengbach catchment (France) indicated seasonal variability of riverine $\delta^{44/42}$Ca, attributed to seasonal input of biologically affected soil porewater [Schmitt et al., 2003a]. A more detailed study of water and vegetation in the same catchment confirmed the seasonal variability in stream $\delta^{44/42}$Ca [Cenki-Tok et al., 2009]. The analysis of Ca isotopes in several Himalayan rivers revealed that rivers had a heavier Ca isotopic composition than the local bedrock due to the precipitation of secondary carbonates [Tipper et al., 2006b, 2008a]. However, despite the range in rock $\delta^{44/42}$Ca values and vegetation differences between sampled rivers, there was little variability in the observed $\delta^{44/42}$Ca values between rivers and only a hint of seasonal variations. An expanded data set of world rivers [Tipper et al., 2010] showed systematic seasonal variations in $\delta^{44/42}$Ca between wet and dry seasons and confirmed the limited variability of $\delta^{44/42}$Ca in large rivers. The first published Ca isotope analysis of soil was from the Santa Cruz chronosequence in California [Bullen et al., 2004]. A sequential extraction procedure was performed on four soils of different ages, and porewater samples were collected. To explain their results the authors postulated that three different soil Ca pools must exist. The amount of seasalt, the cation exchange capacity of soils and organically-complexed Ca were identified as important factors influencing the $\delta^{44/42}$Ca value of the soil pools. A second chronosequence study from Hawaii also observed that the Ca isotopic composition of the soil changed with age, and was influenced by dust deposition and uptake of Ca by vegetation [Wiegand et al., 2005]. Calcium isotopes have been recently used to investigate soil-tree systems [Perakis et al., 2006, Page et al., 2008, Holmden and Bélanger, 2010]. These studies demonstrated the influence of trees on the local soil profile using soil sequential extraction procedures. Significant within-tree Ca isotope fractionation has been reported with an increase in $\delta^{44/42}$Ca values from roots to leaves. Even in soil profiles where there is negligible biological activity, systematic trends of Ca isotopes in soil calcium sulphate and calcium carbonate have been observed with depth [Ewing et al., 2008]. The location of this study was the Atacama desert and Ca isotope fractionation was induced by periodic rain events.

1.3 Research objectives and outline of the thesis

This thesis covers two main areas: glacial hydrology and calcium isotope fractionation in a small catchment. The selected fieldsite was the Damma glacier forefield in central Switzerland (described in chapter 2). Due to the BigLink project, an inter-disciplinary network of scientists, the Damma catchment has been very well characterised. This level of detail allows a greater understanding of the system as a whole.

The aim of the first part of the thesis was to investigate the processes responsible for the seasonal variability in stream water chemistry and assess the impact of seasonal variability on the calculation of chemical weathering fluxes. The data collected from the Damma catchment were then compared to other catchments to investigate whether the glacier had an effect on annual chemical weathering fluxes as compared to similar non-glaciated catchments.

The results of seasonal and diurnal stream water chemistry variations, including major element, ^{87}Sr/^{86}Sr and δ^{18}O measurements, are reported in chapter 3. Processes controlling the stream water chemistry are discussed with reference to the changing hydrology of the catchment over the year.

The following chapter (chapter 4) places the calculated weathering fluxes from the Damma catchment in a global context, specifically focussing on the influence of glaciers on weathering rates and stream water composition by examining a compilation of data from glaciated and non-glaciated granitic and basaltic catchments.

The objectives of the second part of the thesis were to investigate calcium isotope fractionation in a terrestrial environment and to considerably expand the current data set of calcium isotope measurements in terrestrial materials. Specifically, the aim was to investigate whether calcium isotope fractionation was observable at the field scale and thus be a potentially useful tracer of weathering and related biogeochemical processes. To achieve this, many different components of the system, from rocks to plants, were analysed. The Damma catchment was particularly suited to this study since it is a mono-lithological fieldsite with minimal vegetation cover which minimised the number of potential fractionation processes occurring.

The calcium isotopic compositions of soils, rocks and stream water are presented in chapter 5, in addition to data from soil sequential extractions. The Ca isotopic compositions of the different samples are compared in order to identify potential isotope fractionation processes. The effect of these fractionation processes on the stream water chemistry exiting the catchment is explored.

The final chapter (chapter 6) builds on chapter 5 and considers in greater depth the Ca isotope fractionation occurring between soil and plants and between different plant tissues. Data were collected from a range of plant species, allowing inter-species differences in calcium isotope fractionation to be investigated.

Chapter 2

Field area - Damma Glacier

The Damma glacier is a small (10.7 km^2), glaciated catchment situated in the central Swiss Alps near the village of Göschenen in Canton Uri. This field area is the focus of multi-disciplinary scientific research (BigLink - http://www.cces.ethz.ch/projects/clench/BigLink) and is a designated critical zone observatory (CZO). BigLink stands for **Bi**osphere-**Ge**osphere interactions: **Link**ing climate change, weathering, soil formation and ecosystem evolution. There are five main research themes: organic carbon fluxes, microorganisms, plant succession, isotope geochemistry and hydrology. The overall goal is to understand processes occurring during initial soil formation and the links between biology, chemistry, hydrology and geology.

2.1 Glacier history

The Damma glacier is a temperate alpine glacier and the extent of the glacier has been annually monitored since 1921 by the Cryospheric Commission of the Swiss Academy of Science. A comparison between the Siegfried map, which was developed from 1870 to 1922, and the modern topographical map clearly shows the retreat of the former Wintergletscher over the last century (Fig. 2.1). Swiss glaciers grew throughout the Little Ice Age and subsequently retreated after 1850. The large side moraines found in many alpine valleys, including Damma, date from this time. Before 1921 the extent of the glacier can be inferred by comparison with neighbouring glaciers which have longer historical records such as Chelengletscher and Rhonegletscher. These glaciers exhibited a short period of re-advance in the early 20th century which ended in 1921 at Damma and is marked by a terminal moraine. There was another period of re-advance lasting two decades from 1970 to 1990 which formed a second terminal moraine. In the period 1921 to 2003 the cumulative length change of the glacier was -379 m [VAW, 2005]. In 2003, the retreating glacier split into two. A piece of 'dead ice' ($\sim$0.14 km^2) currently lies in the valley floor separated from the main glacier by steep cliffs. The glacier currently covers 40% of the catchment area of 10.7 km^2.

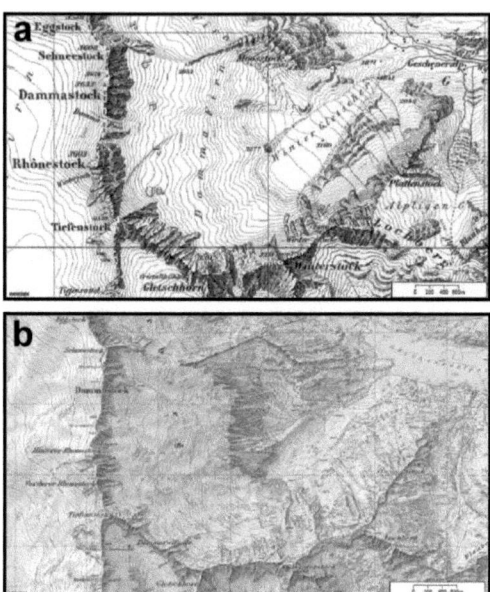

Figure 2.1: A comparison between the Siegfried map (end of 19th century) and the modern topographical map clearly shows the retreat of the former Wintergletscher. The village of Göscheneralp (a) was relocated following the formation of the Göscheneralpsee reservoir (b) in 1960. Maps were sourced from the Swiss Federal Office of Topography.

2.2 Geology

The Damma glacier forefield is situated in the Cental Aar granite zone of the Aar Massif (Fig. 2.2). The Aar Massif is part of the Helvetic crystalline basement of the Alps which has a complex history [Labhart, 2009]. The last major thermal event of the region was the Variscan orogeny when the basement was intruded forming distinct magmatic zones. The dominant geochemical composition is represented by the Central Aar granite, a calc-alkaline granitic suite of rocks [Schaltegger, 1994]. The Central Aar intrusion has been dated at 298±2 Ma [Schaltegger, 1994]. During Alpine orogeny the Aar Massif was lightly metamorphosed between 20 and 65 Ma (greenschist facies), with the intensity of deformation increasing from north to south [Dempster, 1986]. In the present day, the Central Aar granite forms the largest granitic body in Switzerland, outcropping over 500 km^2 [Labhart, 2009]. The elemental compositions of seven rock samples collected from the Damma forefield are reported in de Souza et al. [2010, EA 1-3]. The average rock composition is quartz (32%), plagioclase (32%, An$_{0.1}$), microcline (23%), muscovite (5.6%), biotite (3.6%), epidote (2.8%), together with trace amounts of apatite. Epidote is finely disseminated throughout the rock, similar to other granitic catchments [Mavris et al., 2010, Oliva et al., 2004]. Biotite has been extensively altered to chlorite and negligible levels of carbonate (91 ± 83 mg/kg, $1\sigma_{SD}$) were detected by coulometric titration in the rock samples which have been analysed so far [de Souza et al., 2010].

Figure 2.2: Simplified geological map of Switzerland, highlighting the extent of the Aar Massif and the Variscan intrusions. The fieldsite is situated in the Central Aar granite region.

2.3 Soils and vegetation

Soil samples were collected in September 2007 as part of the BigLink project [Bernasconi et al., 2008]. Twenty-one sampling locations (Soils BL1 - BL21) were chosen by randomly selecting sites within a pre-determined grid. This means that the approximate soil age was pre-determined but the final location was random, allowing spatial heterogeneity to be accounted for whilst covering the length of the chronosequence. In addition, two older sites which were unaffected by glaciation in the last 150 years (reference sites) were picked which were just outside the forefield (Soils BL23 and BL24). The location of the soil samples is shown in Fig. 2.3. This thesis also includes analyses of samples which were taken in 2006 [de Souza et al., 2010, Kiczka-Cyriac, 2010]. These samples also cover the length of the chronosequence but were not randomly distributed and were sampled according to soil horizon.

The ice-free age of the soils (exposure since deglaciation) is known from the historical records of glacial length. The soil chronosequence presently covers approximately 150 years and is around 1.5 km in length. For simplicity, it is assumed that the age of the soil upon exposure is 0, i.e. there is no relict soil from previous glacial retreats, no weathering of the rock underneath the glacier and no input of material of a different age (e.g. from avalanches). The age of soils given in this thesis are the soil ages in 2008.

There are two discontinuities in the chronosequence due to the periods when the glacier advanced, resulting in three distinct sections of development (Fig. 2.4). The

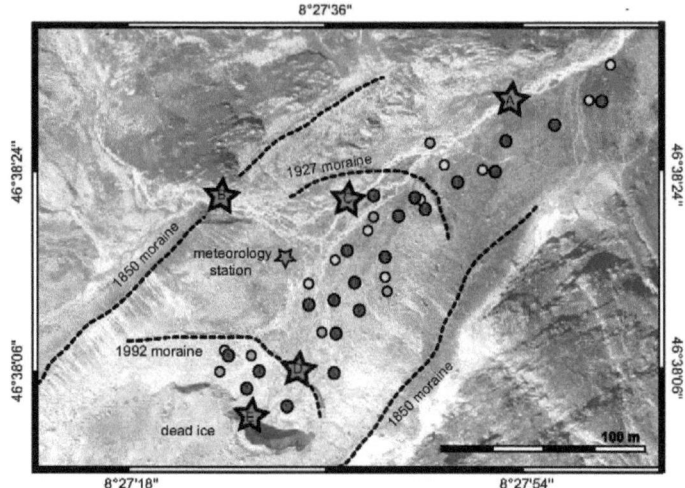

Figure 2.3: Map of the sampling locations. Soil sampling locations are marked by red dots, porewater sampling locations by yellow dots and groundwater sampling locations by orange dots. Soil sampling sites are numbered such that lower numbers are closer to the glacier front and are therefore younger. Two additional soil sampling locations were sampled outwith the forefield (reference sites) and are not shown on the map. Stream water sampling locations are marked by blue stars and the gauging station is located at Site A. Only data from sites A, B and E will be fully discussed. The meteorology station is marked with the green star. The side moraines, which date from 1850, are marked with dashed lines, as are the two most recent terminal moraines.

youngest part (exposed since 1992) is characterised by large boulders and glacial debris and only a few pioneer plants e.g. *Leucanthemopsis alpina*. In the middle section (exposed between 1927 and 1950), soils begin to form and plant cover increases dramatically. The plants are typically herbs and grasses (e.g. *Agrostis gigantea* and *Rumex scutatus*). The oldest section (exposed before the early 1900s) is characterised by the dominance of woody vegetation such as *Rhododendron ferrugineum*.

Figure 2.4: Photographs illustrating the development of vegetation along the chronosequence from (a) Site BL2, 7 years old to (b) Site BL10, 68 years old to (c) Site BL17, 111 years old.

Soils in the chronosequence overlie bouldery morainal material, are poorly weathered, thin and contain large rock fragments. Soils BL1 - BL21 were classified as Hyperskeletic Leptosols (Dystric or Eutric) according to the World Reference Base for Soil Resources [WRB, 2006]. The transition from Eutric (base saturation > 50 %) to Dystric (base saturation < 50 %) occurs after ~ 70 years. Soils BL23 and BL24 were classified as Haplic Cambisols (Dystric, Humic, Skeletic) [WRB, 2006].

2.4 Hydrology and Meteorology

The elevation of the catchment ranges from 1800 m to 3600 m. The average annual temperature is 2.2 °C, annual precipitation is ~2300 mm and annual runoff is ~2700 mm. Evapotranspiration was estimated to be 70 mm in 2008 [Kormann, 2009]. The positive water balance is a result of the retreating glacier. The forefield area is typically covered in snow for around 6 months of the year. A gauging station (Fig. 2.3) recorded water discharge and conductivity every 10 minutes beginning early in the 2008 melt season (Fig. 2.5). In addition to the gauging station, a meteorology station was erected in the middle of the forefield which recorded data every 30 minutes. Rainfall and temperature during the sampling period are shown in Fig. 2.5. Figure 2.6 illustrates the field conditions during each of the water sampling trips and highlights the retreat of snow and the change in vegetation throughout the summer.

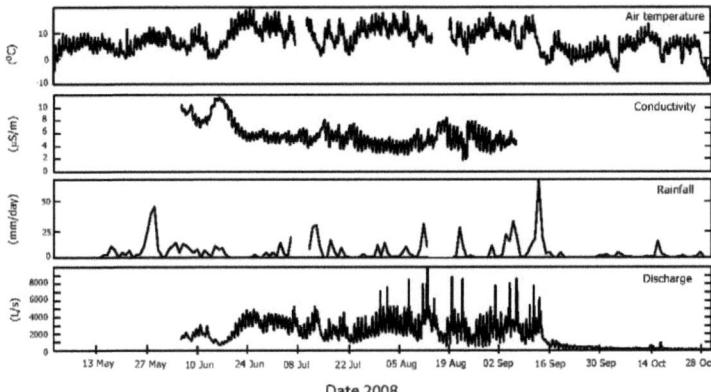

Figure 2.5: Main parameters recorded at the gauging station and at the meteorology station during 2008. Conductivity is inversely related to discharge but does not react strongly to rainfall events which cause large, short-lived spikes in discharge. No data was recorded by the meteorology station from 7th-10th July and 14th-19th August due to technical problems.

The glacier has two sub-catchments. The north-facing part of the glacier drains under the dead ice and forms the main stream, the Dammareuss, which flows through the soil chronosequence. The east-facing part of the glacier drains into a side stream which cuts

Figure 2.6: Photographs taken from the bridge across the river (just above site A) at the start of each water sampling trip. The first photograph was taken on 13th May (a) and the last on 28th October (l). Photographs were taken every two weeks apart from the 22nd July.

through the western side moraine to join the main stream. The confluence of the two streams is just above the 1927 terminal moraine (Fig. 2.3) and the side stream contributes approximately one third to the total discharge. The streams are braided, and there is evidence that the channel positions vary slightly from year to year. Above-surface stream flow ceases during winter.

Water sampling was conducted at fortnightly intervals during the summer of 2008, from 13th May to 28th October. In addition, two winter samples were collected on 16th January and on 8th April 2009. Five locations were sampled in order to assess spatial variability (Fig. 2.7, Table 2.1). The following provides a brief description of the three sites discussed in this thesis: Site E was the stream exiting from under the dead ice, site B was the side stream where it incised the western side moraine and site A was by the gauging station (Fig. 2.3), by which point above-ground braiding had ceased. Two further locations, sites C and D, were sampled but the data from these two sites is not discussed further, however, the data can be found in Table B.1. Two 24-hour sampling campaigns

Table 2.1: Coordinates of water sampling locations (WGS 84 grid)

Site	Latitude	Longitude	Altitude (masl)
A	46°38.468	08°27.991	1963
B	46°38.357	08°27.534	2102
C	46°38.372	08°27.720	1941
D	46°38.186	08°27.638	2050
E	46°38.121	08°27.554	2060

with hourly sampling were conducted on 24th-25th August 2008 and 2nd-3rd June 2009 at Site A.

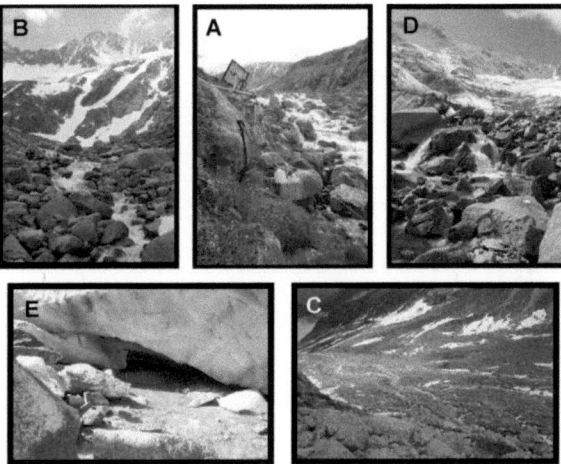

Figure 2.7: Photographs of the five water sampling locations. See Figure 2.3 for exact locations. The photograph 'C' is taken from site B, the actual sampling location of C is the confluence of the side stream with the nearest branch of the main stream (approximately middle of the picture).

Chapter 3

Hydrological control of stream water chemistry in a glaciated catchment (Damma Glacier, Switzerland)*

3.1 Introduction

Globally, the chemical composition of river water varies widely [Meybeck, 2003]. Although much of this variation can be attributed to different lithologies [Reeder et al., 1972, Gaillardet et al., 1999], there is considerable variation between rivers draining similar lithologies [Gíslason et al., 1996, White et al., 1999b, Oliva et al., 2003]. The relative importance of the different extrinsic weathering factors such as climate, runoff, tectonics and vegetation on runoff composition is still debated [Bluth and Kump, 1994, White and Blum, 1995, Oliva et al., 2003, Riebe et al., 2004, West et al., 2005] and disentangling these factors is difficult since they are not independent variables, e.g. temperature affects runoff and local vegetation. The majority of studies that have derived estimates of CO_2 consumption by silicate weathering, based on river chemistry, have assumed that the chemical composition of a single river derived from spot sampling is invariant. However, an increasing number of studies of large rivers indicate significant seasonal variations in chemical composition [Cameron et al., 1995, Yang et al., 1996, Shiller, 1997, Galy and France-Lanord, 1999, France-Lanord et al., 2003, Zakharova et al., 2005, Tipper et al., 2006a, Ollivier et al., 2010] and long term studies highlight the inter-annual variability of annual weathering fluxes [Likens et al., 1998, Gupta et al., 2011]. Time series data are vital in order to obtain accurate annual chemical weathering fluxes but also offer an unrivalled opportunity to understand the underlying

---

*A modified version of this chapter has been published in *Chemical Geology*: R.S. Hindshaw, E.T. Tipper, B.C. Reynolds, E. Lemarchand, J.G. Wiederhold, J. Magnusson, S.M. Bernasconi, R. Kretzschmar and B. Bourdon. Hydrological control of stream water chemistry in a glacial catchment (Damma Glacier, Switzerland). *Chem. Geol.*, 285: 215-230, 2011.

mineral weathering processes. This is because lithology, the major control of stream water chemistry, is constant whilst external variables such as climate and hydrology vary, permitting an assessment of these variables on chemical weathering processes.

High frequency sampling studies of rivers to investigate temporal variability have tended to focus on alpine catchments which are either glaciated [Fairchild et al., 1999, Hodson et al., 2000, Mitchell et al., 2001, Hosein et al., 2004] or snow covered [e.g. Marsh and Pomeroy, 1999]. The bias towards alpine catchments results from these catchments' importance in hydroelectric power and summer water supply [Viviroli and Weingartner, 2004]. Consequently, the focus of many of these studies has been to ascertain where the *water* comes from and not where the *solutes* originated [Malard et al., 1999]. As a result, the hydrology of glacial catchments tends to be very well understood making them an ideal choice for gaining a better understanding of chemical weathering processes.

Combining the hydrological approaches (e.g. hydrograph separation [Buttle, 1994]) with the chemical approaches (e.g. mass balance [Garrels and Mackenzie, 1967]) offers a unique opportunity to improve our understanding of the processes which control stream water chemistry. A number of studies have used such a combination, most notably that of Clow and Drever [1996] who artificially enhanced discharge and studied the resultant effects, proving that discharge has an important control over solute acquisition. Other studies have found the role of discharge to be less important, invoking chemostatic behaviour [Godsey et al., 2009, Clow and Mast, 2010]. These studies strongly indicate that although the weathering of primary minerals provides solutes, it is the secondary processes such as adsorption [Berner et al., 1998] and secondary mineral formation which actually control the resultant chemical composition of the stream. These processes are not constant over the year [Clow and Mast, 2010] and are very difficult to constrain. More data is needed to assess their relative importance.

Glacial catchments are ideal for exploring and further constraining the controls on stream water chemistry. Due to the large fluctuations in discharge over diurnal and seasonal timescales, the role of discharge in changing stream water chemical compositions can be evaluated. In addition, the changes in glacier hydrology over the season and the development of glacial drainage systems are relatively well documented [e.g. Brown, 2002]. Glacial studies are typically only conducted during the melt season, consequently, potential chemical changes from summer to winter are not well known. Additionally, the role of glaciers in enhancing or suppressing chemical weathering rates is still contested [Sharp et al., 1995, Anderson et al., 1997a].

Studies which use the dissolved load composition in order to understand weathering reactions often have weak links to catchment hydrology. In this study, we investigate whether hydrological source information can be used to explain the observed daily and seasonal changes occurring in glacial stream water chemistry in a small, granitic, alpine catchment by utilising major element, $^{87}Sr/^{86}Sr$ and $\delta^{18}O$ data. Identifying the main controls of solute acquisition will improve the quantification of weathering rates in catchments where discharge data is available but frequent chemical sampling is not possible.

3.2 Methods

This chapter discusses data from three of the water sampling locations: sites A, B and E (Fig. 2.3). The location of these sites and the sampling campaign are described in section 2.4 and the collection of samples is described in appendix A.1.2. The water samples were analysed for major anions, cations and silicon (Appendix A.2.1), Sr isotopes (Appendix A.5) and O isotopes (Appendix A.6).

3.3 Hydrology and Water Chemistry

3.3.1 Catchment meteorology and hydrology

Air temperature, conductivity, rainfall and discharge all showed significant temporal variability during 2008 (Fig. 2.5). As expected for a glacial catchment, discharge exhibited a strong seasonal trend with maximum discharge during the summer months and an abrupt cessation of flow coinciding with the first snow in autumn. Heavy rain events in the summer caused rapid increases in discharge with flows reaching in excess of 6000 L s^{-1}. Superimposed on the seasonal discharge trend were strong diurnal cycles linked to the melting of the glacier during the day. Conductivity was inversely related to discharge and exhibited diurnal and seasonal variations. The conductivity of the stream water was not strongly correlated with heavy rainfall events.

Pronounced seasonal and diurnal variations were observed in $\delta^{18}O$ which were similar in magnitude at all three sites (Fig. 3.1). Stream water sampled at the start of the melt season was depleted in ^{18}O ($\delta^{18}O$ ~-17‰) and as melting progressed the $\delta^{18}O$ value measured in the stream rapidly increased. From July onwards $\delta^{18}O$ continued to increase, but at a slower rate, until the winter months when no further increase in $\delta^{18}O$ was observed. The highest $\delta^{18}O$ values reached were around -14‰. Similar seasonal trends in $\delta^{18}O$ have been previously recorded in snow affected catchments [Bottomley et al., 1986, Taylor et al., 2001, Unnikrishna et al., 2002, Liu et al., 2004].

The diurnal amplitudes of $\delta^{18}O$ variation were 0.3‰ and 0.6‰ in June and August respectively, which were less than the seasonal amplitude of 3‰ (Tables 3.1 and 3.2). In June, $\delta^{18}O$ was inversely related to discharge but there was a significant phase shift with the minimum $\delta^{18}O$ value reached seven hours after maximum discharge. The diurnal minimum can be explained by increased melting in response to increased solar radiation during the day. The first half of the August $\delta^{18}O$ diurnal cycle was also inversely related to discharge, but the minimum $\delta^{18}O$ value was observed two hours before maximum discharge. In August a second minimum occurred during the night and this could have been caused by ice melt contributions at night due to high night-time air temperatures [Jobard and Dzikowski, 2006].

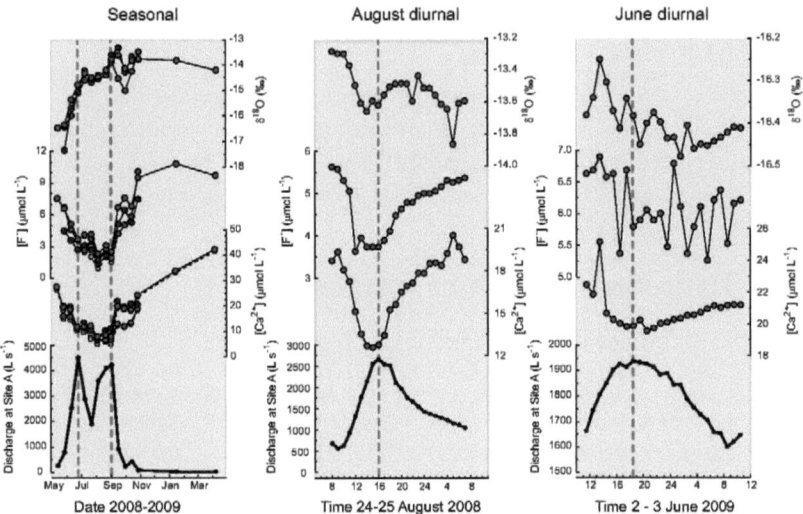

Figure 3.1: Left column: Seasonal changes in discharge, cation concentrations (represented by Ca^{2+}), anion concentrations (represented by F^-) and $\delta^{18}O$. Red symbols are data from Site A, green symbols are for Site B and blue symbols are for Site E. Precipitation corrected values are indicated by orange, light green and light blue for Sites A, B and E respectively and are joined by a dashed line. For most data points the correction is too small to observe. Centre and right columns: Diurnal data for discharge, cation concentrations, anion concentrations and $\delta^{18}O$, from site A during August and June respectively. The vertical dashed lines indicate times of maximum discharge.

3.3.2 Precipitation

There are three sources of precipitation inputs to the catchment to consider: rain, snow and indirectly ice, each of which has a different chemical and isotopic composition [Table 3.1, Tresch, 2007].

The $\delta^{18}O$ values for rain decreased from -7‰ in spring to -14‰ in late September. The decrease in $\delta^{18}O$ throughout the season is caused by changes in the precipitation source characteristics and air temperature [Gat, 1996, Unnikrishna et al., 2002]. The chemical composition of the rain was variable and was dominated by Ca^{2+} and NO_3^-. The average $^{87}Sr/^{86}Sr$ of two composite rain samples was 0.70941 which is similar to the $^{87}Sr/^{86}Sr$ of rain recorded in a neighbouring catchment [Arn et al., 2003] and is considerably lower than the ratios measured in the stream waters (0.71570 to 0.73248, see section 3.3.4).

The $\delta^{18}O$ values in snow were lower than for rain and were variable with time, depth

Hydrology and Water Chemistry

Table 3.1: Major species, $^{87}Sr/^{86}Sr$, and $\delta^{18}O$ data for seasonal sampling. R=rain and S=snow. Measurement reproducibility is described in the text.

Site	Date (YMD)	Time (CET)	Runoff (L s^{-1})	pH	T (°C)	Ca^{2+}	Mg^{2+}	Na$^+$	K$^+$	Si	F$^-$	Cl$^-$	NO$_3^-$	SO$_4^{2-}$	HCO$_3^-$	Sr^{2+} (nmol L^{-1})	$\delta^{18}O$ ‰	$^{87}Sr/^{86}Sr$
										(μmol L^{-1})								
A	20080513	07:35	250	5.61	0.8	27.5	5.0	13.2	14.5	26.2	7.5	5.2	34.0	8.9	63.1	36.6	-16.47	0.72680
A	20080527	15:20	800	6.55	0.4	19.5	3.1	9.8	10.2	15.8	6.6	4.1	22.2	6.8		27.8	-16.36	0.72458
A	20080610	15:30	2528	6.34	4.7	18.7	3.3	8.9	9.5	14.6	5.1	3.6	15.8	5.9	26.2	28.3	-15.37	0.72070
A	20080624	15:30	4502	6.42	5.5	11.8	2.0	7.1	7.6	7.7	3.7	3.6	9.4	3.9	18.1	17.8	-14.78	0.72064
A	20080708	14:40	2886	6.23	6.3	11.8	2.1	6.3	7.9	9.9	4.1	3.0	9.1	4.0	16.6	18.1	-14.33	0.72194
A	20080722	15:30	1908	6.24	5.4	12.2	2.4	6.7	8.1	12.2	4.2	2.4	8.5	4.3	32.8	19.2	-14.50	0.72268
A	20080805	14:35	3595	6.51	10.5	7.9	1.5	4.3	5.5	6.4	2.3	1.9	5.2	2.5	13.9	12.3	-14.47	0.72310
A	20080819	15:55	4110	6.52	6.9	12.2	1.5	6.7	6.2	7.3	2.3	2.9	4.7	2.7	16.1	12.7	-14.19	0.72440
A	20080902	13:55	4221	6.15	5.9	8.5	1.5	5.8	4.7	6.1	3.1	1.9	4.8	2.6	13.4	11.7	-13.78	0.72486
A	20080916	15:30	916	5.84	4.9	7.9	1.4	3.3	16.1	30.0	2.5	2.9	14.5	9.1	40.6	36.1	-13.63	0.72278
A	20080930	17:00	236	5.75	4.2	19.0	4.1	13.9	16.1	37.6	6.7	3.2	12.7	9.1	49.6	34.0	-14.14	0.72319
A	20081014	14:30	420	5.75	4.2	19.1	4.5	15.5	13.2	24.1	7.6	2.5	16.6	7.9	33.6	34.7	-13.83	0.72320
A	20081028	17:00	84	5.55	2.6	18.6	3.6	12.4	14.7	54.3	6.8	2.5	14.7	7.9	97.5	42.7	-13.76	0.72489
A	20090116	09:50	20	6.30	2.2	24.3	5.5	24.3	9.5	71.1	10.8	2.6	15.7	10.3	80.4	48.2	-13.83	0.72370
A	20090408	08:05	20	6.05	2.5	33.9	7.8	29.8	24.9	40.9	9.7	2.7		14.3	72.6	73.7	-14.20	0.72445
B	20080527	13:50		6.37	-0.2	16.0	2.2	7.3	6.8	70.6	6.7	4.7	20.2	6.0		17.1	-16.43	0.73228
B	20080610	12:50		6.15	3.8	15.7	2.1	6.8	6.8	9.7	4.6	5.0	15.6	4.5		18.0	-15.99	0.72582
B	20080624	12:45		5.34	7.0	11.0	1.5	3.8	3.8	5.0	3.6	2.6	10.4	3.9	7.8	13.4	-15.10	0.72648
B	20080708	12:40		5.95	8.6	11.3	1.5	4.8	5.3	3.3	2.6	2.6	12.2	3.9	9.9	14.4	-14.60	0.72663
B	20080722	11:00	421	5.28	8.2	11.0	1.6	4.5	4.8	4.5	3.7	4.1	10.6	4.0	6.7	13.7	-14.69	0.72902
B	20080805	12:25	2092	6.43	8.4	6.5	0.9	1.7	2.4	2.0	1.7	1.0	4.4	1.8	20.7	7.7	-14.55	0.72975
B	20080819	13:00	2045	5.50	10.5	6.6	0.9	2.2	3.3	2.5	2.8	1.4	3.8	1.8	5.8	7.6	-14.40	0.73166
B	20080902	11:40	2078	5.97	7.4	6.7	0.9	2.2	3.6	2.6	1.7	1.7	4.7	2.0	7.8	7.8	-13.87	0.73248
B	20080916	13:10	175	5.88	6.7	12.9	1.7	4.7	4.9	7.3	4.2	3.0	14.5	5.8	6.0	17.6	-14.54	0.73140
B	20080930	12:15		5.90	4.5	12.0	1.6	5.5	5.5	12.3	5.1	1.7	19.6	3.8	16.6	15.1	-15.01	0.73095
B	20081014	12:10		4.90	6.1	13.4	1.8	6.0	6.4	11.2	5.8	2.2	19.0	6.0	12.8	19.0	-14.25	0.73063
B	20081028	10:20		7.30	3.5	19.0	3.0	16.0	11.0	39.3	10.1	2.8	15.9	6.6	12.0	24.3	-13.48	0.73018
B	20090527	09:30		5.83	0.2	20.0	2.9	8.9	10.0	9.4	4.5	6.5	28.9	8.4	69.8	38.6	-17.34	0.71875
E	20080610	09:30		6.40	1.3	19.7	3.7	9.3	11.3	10.3	3.6	5.4	17.8	6.8	30.6	37.7	-15.72	0.71570
E	20080624	09:15		7.19	2.5	13.0	2.1	6.6	8.6	10.3	5.4	5.4	10.7	3.0	17.1	20.3	-15.05	0.71712
E	20080708	11:10		7.19	1.1	13.4	2.1	7.1	9.3	6.7	3.1	3.6	12.0	3.0	22.1	24.1	-14.22	0.71761
E	20080722	14:00	892	6.10	2.0	10.3	2.2	6.9	8.2	7.9	2.7	4.1	7.8	3.8	28.1	20.1	-14.58	0.71820
E	20080805	09:00	1866	6.10	3.2	7.4	2.1	4.6	7.1	5.7	1.3	4.3	4.9	2.2	5.3	13.1	-14.42	0.71835
E	20080819	09:30	1083	6.20	3.1	10.3	2.2	5.7	-	8.9	2.5	2.6	6.1	3.9	20.4	19.5	-14.25	0.71889
E	20080902	08:40	723	5.98	2.5	11.0	2.2	5.3	7.8	8.1	2.2	2.4	7.5	4.5	17.1	22.6	-13.62	0.71892
E	20080916	09:45	230	6.58	1.1	21.9	5.3	13.9	17.9	26.9	2.5	2.7	17.1	12.2	50.8	50.9	-13.32	0.71831
E	20080930	09:30	50	6.16	1.2	19.7	4.9	14.0	17.7	31.9	6.4	2.9	17.3	8.9	46.4	40.9	-14.29	0.71946
E	20081014	09:00		6.45	1.5	21.0	3.8	10.9	14.4	19.3	5.3	2.2	20.8	9.0	37.8	45.7	-13.67	0.71938
E	20081028	12:40		5.85	1.3	20.7	5.4	16.2	19.6	35.2	7.5	2.2	20.5	10.8	54.7	45.2	-13.79	0.71990
R	20080624					4.6	0.5	2.3	2.0	2.4	0.0	4.9	22.4	5.4		2.8	-8.57	
R	20080708					12.7	1.4	2.7	2.3	1.6	0.0	4.7	20.8	6.9		13.5	-6.41	
R	20080726					5.2	0.6	2.5	2.2	1.6	0.0	3.0	12.4	3.8		3.7	-10.09	
R	20080805					7.8	1.2	4.2	2.5	0.1	0.0	5.8	23.2	8.0		10.3	-7.08	0.70933[1]
R	20080819					7.1	0.8	2.3	1.7	2.3	0.0	3.2	11.9	4.4		6.0	-8.55	
R	20080902					5.6	0.6	1.1	0.8	1.8	0.0	2.3	17.5	6.5		4.7	-7.40	
R	20080916					16.5	1.3	3.7	4.9	0.9	0.0	3.0	11.5	5.5		18.9	-10.54	
R	20080930					19.2	4.9	6.0	7.5	11.9	0.0	8.1	16.8	5.5		18.8	-13.96	
S	20080513					2.2	0.1	1.7	0.7	0.05	0.0	6.4	4.9	1.0		1.7	-16.38	0.70950[2]
S	20081028					6.2	0.4	4.0	1.8	0.00	0.0	21.0	15.5	3.2		2.9	-10.85	
S	20090116					3.8	0.2	1.4	8.8	0.00	0.0	20.4	8.8	3.2		1.8	-14.90	
S	20090116					1.8	0.2	2.0	2.3	0.01	0.0	6.7	6.9	1.7		1.6	-18.29	
S	20090406					2.7	0.4	3.8	26.6	0.00	0.4	7.4		2.2		2.9	-16.76	
S	20090407					4.9	0.8	0.9	1.6	0.11	0.0	29.9	9.8			6.4	-15.37	
S	20090407					2.5	0.2	0.9	0.9	0.13	0.0	9.3		0.6		1.1	-14.12	
S						0.6	0.1	0.4	0.6	0.09	0.0	4.0	6.1			0.5	-15.10	

Discharge values in italics were estimated.
PO$_4^{3-}$ was measured but was below the detection limit
[1] Average rain, first 4 samples of the season
[2] Average rain, last 4 samples of the season

Table 3.2: Major species, $^{87}Sr/^{86}Sr$ and $\delta^{18}O$ data for diurnal sampling. Measurement reproducibility is described in the text.

Time (CET)	Discharge (L s^{-1})	pH	T (°C)	Ca^{2+}	Mg^{2+}	Na$^+$	K$^+$	Si	F$^-$	Cl$^-$	NO$_3^-$	SO$_4^{2-}$	HCO$_3^-$	Sr^{2+} (nmol L^{-1})	$\delta^{18}O$ ‰	$^{87}Sr/^{86}Sr$
24-25 August 2008							(μmol L^{-1})									
08:00	677	5.78	3.7	18.7	3.7	10.5	13.2	20.5	5.6	2.7	11.6	7.2	29.2	28.2	-13.28	0.72289
09:00	559	5.52	4.7	19.3	3.8	10.9	12.5	22.0	5.6	2.5	11.6	7.7	32.4	29.5	-13.30	
10:00	625	5.61	5.3	18.1	3.5	9.8	11.9	19.0	5.3	2.1	11.5	7.3	31.1	26.9	-13.30	
11:00	931	5.52	6.0	17.3	3.3	9.4	11.4	17.6	5.1	1.9	10.5	6.6	27.8	26.5	-13.37	0.72229
12:00	1325	5.67	6.5	15.1	2.9	7.6	9.9	15.0	3.6	1.8	9.4	5.5	26.3	22.7	-13.50	
13:00	1766	5.61	6.8	13.5	2.6	6.7	8.8	13.2	3.9	1.6	8.2	4.8	22.1	20.5	-13.61	
14:00	2152	5.71	6.8	12.7	2.7	6.4	9.1	12.7	3.7	1.7	7.8	4.6	20.8	19.5	-13.66	0.72290
15:00	2557	5.73	6.6	12.6	2.3	5.8	7.9	11.4	3.7	1.5	7.5	4.3	19.9	19.0	-13.59	
16:00	2666	5.53	6.1	12.8	2.4	6.1	8.3	12.1	3.7	1.3	7.6	4.4	20.9	19.6	-13.62	
17:00	2542	5.55	5.5	13.4	2.5	6.4	8.2	12.6	3.9	1.3	8.0	4.8	23.9	20.7	-13.56	0.72324
18:00	2514	5.48	4.7	15.3	3.0	7.3	9.2	15.8	4.1	1.3	8.6	5.1	26.3	18.5	-13.50	
19:00	2108	5.70	4.6	15.6	3.0	8.0	10.1	16.7	4.5	1.4	9.0	5.5	26.2	23.5	-13.49	
20:00	1961	5.54	4.3	16.5	3.1	8.4	10.1	17.6	4.6	1.4	9.5	5.8	26.1	23.8	-13.48	0.72311
21:00	1755	5.70	4.1	16.9	3.3	8.9	10.6	19.1	4.8	1.5	9.9	6.1	28.0	23.8	-13.49	
22:00	1658	5.80	4.0	17.2	3.2	9.0	10.5	18.7	4.8	1.4	10.3	6.1	28.1	25.4	-13.60	
23:00	1539	5.67	3.9	17.9	3.5	9.5	11.1	20.5	4.9	1.6	10.6	6.5	30.5	27.2	-13.44	0.72306
24:00	1427	5.67	3.8	17.8	3.5	9.6	11.1	20.7	5.0	1.6	10.5	6.4	30.5	27.4	-13.51	
01:00	1375	5.72	3.8	18.5	3.7	10.2	11.6	22.1	5.1	1.6	10.8	6.7	35.5	28.4	-13.51	
02:00	1319	5.67	3.8	18.6	3.6	10.2	11.6	22.4	5.2	1.6	11.1	6.7	35.0	25.1	-13.56	0.72296
03:00	1274	5.49	3.8	18.4	3.6	10.1	11.6	21.4	5.2	1.6	11.2	6.7	32.4	24.8	-13.62	
04:00	1226	5.68	3.7	19.2	3.7	10.1	11.9	22.6	5.3	1.8	11.5	6.9	33.2	28.3	-13.64	
05:00	1150	5.75	3.7	20.5	3.7	10.6	11.8	22.0	5.3	1.6	11.2	6.7	34.0	29.1	-13.86	0.72313
06:00	1111	5.69	3.8	19.7	3.8	11.3	12.6	23.1	5.3	1.8	11.8	6.9	34.6	30.0	-13.61	
07:00	1041	5.76	3.9	18.8	3.7	11.1	12.5	22.8	5.4	1.9	11.7	7.0	31.1	28.8	-13.59	
2-3 June 2009																
11:30	1663	7.61	3.6	22.5	5.0	12.5	13.7	12.3	6.6	9.2	23.3	7.8	40.2	30.7	-16.38	
12:30	1744	7.62	3.6	21.9	5.4	11.7	14.2	12.8	6.7	4.7	20.9	7.6	40.5	30.7	-16.34	
13:30	1805	6.86	3.2	25.2	4.0	10.9	12.0	10.9	6.9	3.4	21.3	7.4		29.2	-16.25	
14:30	1850	6.80	2.9	20.7	3.9	10.0	11.2	11.0	6.6	2.5	19.2	7.3	32.1	28.1	-16.30	
15:30	1902	6.82	2.9	20.3	3.8	9.8	11.2	10.2	6.6	2.5	19.3	7.4	32.7	27.1	-16.37	
16:30	1926	6.68	2.7	20.0	4.1	10.7	12.0	10.4	5.4	3.5	19.7	7.0		26.7	-16.41	
17:30	1914	6.77	2.5	19.8	3.7	9.4	10.4	10.1	6.7	2.4	18.4	6.9	30.7	24.8	-16.34	
18:30	1937	6.81	2.3	19.9	4.0	9.5	10.8	10.1	5.8	2.4	18.2	7.0		28.0	-16.38	
19:30	1931	6.83	2.2	20.2	4.8	9.5	11.6	10.1	6.1	2.4	18.8	7.0	31.4	28.9	-16.45	
20:30	1926	6.59	2.1	19.5	3.6	9.2	10.2	11.2	5.8	2.5	18.7	7.2	30.1	27.6	-16.40	
21:30	1914	6.75	2.0	19.7	3.6	9.3	9.7	9.7	6.1	2.4	18.6	7.1		27.5	-16.38	
22:30	1885	6.78	2.0	20.0	3.6	9.4	10.5	9.7	5.9	3.4	19.0	7.1		29.1	-16.40	
23:30	1891	6.77	2.0	20.1	3.6	9.7	10.6	9.8	6.0	3.3	21.8	7.0	33.8	28.0	-16.44	
00:30	1845	6.55	2.0	20.3	3.6	9.7	11.3	9.9	5.5	2.8	21.3	7.2	30.2	26.6	-16.43	
01:30	1845	6.80	1.9	20.4	3.8	9.7	11.1	10.3	6.8	2.8	20.2	8.2		28.3	-16.48	
02:30	1788	6.86	1.9	20.5	3.8	9.8	10.8	10.5	6.1	2.4	19.4	7.2		28.8	-16.48	
03:30	1755	6.82	1.8	20.6	3.8	9.9	11.2	10.3	5.4	3.4	20.0	7.3	33.8	29.3	-16.46	
04:30	1728	6.60	1.7	20.7	3.9	9.6	10.9	10.7	5.8	2.7	19.8	7.7	32.3	28.7	-16.45	
05:30	1706	6.74	1.9	20.5	3.8	10.3	11.1	11.4	5.3	4.0	20.1	7.6		29.1	-16.45	
06:30	1658	6.78	2.1	21.1	4.0	9.7	11.7	11.0	6.2	2.8	20.0	8.0	31.9	27.8	-16.44	
07:30	1653	6.80	2.5	21.0	3.9	9.7	11.8	11.1	6.4	2.8	20.1	9.7	34.2	30.5	-16.43	
08:30	1600	6.76	3.0	21.2	4.0	9.8	11.8	11.3	5.5	3.3	20.2	8.0		31.0	-16.42	
09:30	1621	6.80	3.5	21.2	4.4	10.4	11.8	11.8	6.2	3.0	20.2	7.8		29.1	-16.41	
10:30	1647	6.90	3.9	21.2	4.0	9.9	11.8	11.2	6.2	3.0	19.8	7.8		29.6	-16.41	

Table 3.3: Major species, $^{87}Sr/^{86}Sr$ and $\delta^{18}O$ data for groundwater (GW) and pore water (PW) samples. Measurement reproducibility is described in the text.

Sample name	Ca^{2+}	Mg^{2+}	Na^+	K^+	Si	F^- ($\mu mol\ L^{-1}$)	Cl^-	NO_3^-	PO_4^{3-}	SO_4^{2-}	HCO_3^-	Sr^{2+} ($nmol\ L^{-1}$)	$\delta^{18}O$ ‰	$^{87}Sr/^{86}Sr$
GW1a	23.2	4.8	20.2	17.4	59.1	12.1	5.3	12.1	-	7.8	62.7	34.6	-13.11	0.72959
GW1b	59.0	5.1	17.2	15.5	39.7	15.3	94.7	7.5	-	7.7	36.2	52.1	-14.17	
GW1c	34.8	5.4	16.4	15.6	42.6							36.9		0.72831
GW2a	49.9	16.8	21.7	41.3	45.2	3.6	12.9	27.2	-	5.8		100.1	-13.96	
GW2b	41.2	9.4	22.7	23.2	53.1	9.6	1.6	4.6	-	11.4	127.8	53.5	-14.09	
GW2c	47.5	10.3	24.1	25.2	54.9	9.4	21.7	11.0	-	12.5	106.8	63.9	-14.08	
GW3a	120.5	23.7	22.4	42.6	78.1	6.3	809.4	-	-	3.6		144.8	-13.10	
GW3b	51.2	20.9	20.5	38.0	74.7	4.4	33.1	-	-	2.3	204.5	80.1	-13.27	
GW3c	42.4	19.2	18.4	38.6	62.2							75.5		0.71553
GW4b	13.7	3.0	3.9	13.0	4.3	2.4	2.3	5.4	-	3.4	29.6	21.8	-14.23	
GW5b	13.6	2.4	3.7	12.6	5.4	2.6	1.4	6.9	-	4.2	27.4	21.9	-14.12	
GW6b	14.7	5.9	9.1	19.6	25.8	8.5	1.4	6.8	-	4.2	55.4	18.1	-14.78	
PW1a	152.0	14.2	27.6	44.5	15.7	0.0	468.5	10.6	-	2.5		45.6	-18.18	
PW1b	149.3	49.6	18.8	17.6	17.2	1.7	223.4	-	-	2.1		119.3	-10.29	
PW1c	22.3	8.6	23.1	32.3	61.9	3.4	19.0	1.4	-	8.7	-	30.6	-12.10	0.71691
PW1d	19.7	8.5	22.4	30.2	61.5	3.3	10.2	1.7	-	9.0	-	30.0	-12.24	
PW1e	33.1	13.1	22.8	27.7	42.9	3.5	10.2	-	-	7.6	117.9	46.2	-13.60	
PW2a	2760.8	59.9	19.5	139.5	52.4							1114.0		0.70840
PW2b	75.3	5.8	15.8	44.2	14.9	3.3	250.2	6.5	1.9	1.7		23.2	-14.12	0.71577
PW2c	268.5	149.4	184.0	196.8	15.0	9.3	677.0	836.6	4.6	293.5		272.8	-8.72	0.71570
PW3a	37.0	6.8	1.7	15.7	26.2	4.3	96.9	0.4	-	0.8		27.2		0.71652
PW3b	59.2	15.5	18.6	29.4	9.0							59.8		
PW4	550.4	95.5	47.5	138.5	93.9	15.6	1135.8	-	-	3.3	579.3	493.6	-14.04	0.72276
PW5	591.7	8.2	51.8	26.5	20.5	0.0	3568.1	-	38.5	0.0		40.1	-19.29	
PW6	78.0	10.7	18.7	8.4	34.7	0.9	159.9	-	-	1.7		22.0	-6.42	
PW8	57.7	1.1	352.9	11.0	2.7	2.5	125.5	2.7	-	2.2		23.7	-14.21	
PW11	428.2	66.6	47.4	374.3	89.0	12.5	1304.3	11.1	-	24.9	570.8	279.8	-13.03	0.71447
PW13	3974.5	37.6	33.6	53.9	108.1	21.4	-	24.4	42.0	6.2		2753.8	-9.05	
PW21	339.7	45.0	34.8	102.5	50.3							331.2		0.71237

- indicates species was below the detection limit
blank space indicates species was not measured

Table 3.4: Annual element fluxes

Element	Annual flux (uncorrected)	Annual flux* (kmol/km^2/yr)	Annual flux**	External contribution to annual flux* (%)
Ca	36	32	29	11
Mg	7	6	6	5
Na	21	19	18	10
K	24	21	21	12
Si	34	33	33	1
Cl	9	n.a.	8	-
F	12	11	12	5
NO$_3$	28	21	14	23
SO$_4$	13	11	8	12
HCO$_3$	64	n.d.	n.d.	n.d.
Total cationic flux (meq/m^2/yr)		117	108	

*chloride method
**hydrological method

and altitude (Appendix B), reflecting a heterogeneous snowpack [Unnikrishna et al., 2002]. Snow samples were more dilute than rain samples and the dominant cation and anion was K^+ or Ca^{2+} and Cl^- respectively. Si and F^- concentrations were negligible compared to those measured in the stream. A previous study in the same area reported a $^{87}Sr/^{86}Sr$ value of 0.70985 for snow [de Souza et al., 2010], which is similar to rain.

Ice is directly derived from snow and is also a 'precipitation' source of ions when it melts. The average $\delta^{18}O$ ratio of melted ice from the dead ice was -16.66±0.47‰. No $^{87}Sr/^{86}Sr$ value was obtained for ice. During ice formation, solutes are excluded resulting in very low solute concentrations [Fountain, 1996].

3.3.3 Correction for precipitation inputs

The water chemistry in the Damma catchment is characterised by very dilute meltwaters (sum of positive charge <100 meq L^{-1}), typical of waters draining granitic lithologies [White and Blum, 1995]. The dissolved load is a mixture of precipitation and weathering products, and these dilute stream waters could potentially be strongly influenced by precipitation (including glacial melt).

In order to assess the contribution of weathering to the dissolved load, external inputs need to be corrected for. This is commonly achieved by using a Cl^- correction which assumes that all Cl^- measured has originated from precipitation. However, considering only rain is insufficient in glacial catchments for the following reasons: (1) The majority of precipitation occurs as snow which has a different chemical composition compared to rain, (2) there is a seasonal cycle with increased ice melt in the summer and (3) there is a daily melt cycle where the melting of snow and ice can nearly double the discharge (Table 3.2). Thus, it is necessary to determine the relative contributions of rain, snow and ice melt at the relevant time intervals and measure the chemical composition of each of these water sources in order to assess the contribution of precipitation to the dissolved load. For this study, 5 snow samples, 8 rain samples and two ice samples from 2007 [Tresch, 2007] were collected. The eight rain samples were chemically different from each other (Table 3.1) and the element to chloride ratios did not correlate with the amount of rain which fell in the two week sampling period. The average rain composition was used rather than the rain composition weighted with respect to precipitation volumes due to the frequent overflowing of the bottle collecting the rain. The chemical composition of snow varies during snow melt due to preferential leaching and fractionation of ions [Johannessen and Henriksen, 1978, Tsiouris et al., 1985, Williams and Melack, 1991, Marsh and Pomeroy, 1999] and this will result in a non-constant X/Cl ratio during the snow melt period. Preferential leaching effects are strong for nitrate and sulphate relative to chloride [Williams and Melack, 1991, Marsh and Pomeroy, 1999], thus the precipitation corrections for these two anions may introduce a seasonal bias. Preferential elution of cations relative to chloride is much less pronounced [Marsh and Pomeroy, 1999], thus the precipitation correction for cations will be less affected than nitrate and sulphate by this leaching process. Due to the lack of samples collected during the snow melt period we could not quantify this process further.

The percentage contributions of each of the water sources (rain, snow melt and ice melt) to the total discharge were obtained using the ALPINE3D distributed model for the 2008 hydrological year (Lehning et al. [2006] and applied to the Damma glacier catchment by Magnusson et al. [2011]). ALPINE3D is an energy-balance model which can be used to model high-resolution discharge dynamics in glacio-nival watersheds. The model takes into account the observed evolution of the snow pack during the ablation season [Farinotti et al., 2010] and local meteorological parameters. Snow melt dominated the first half of the summer and ice melt dominated at the end of the summer (Fig. 3.2a). With the modelled contribution of sources known for each sampling day, a weighted X/Cl ratio was calculated and the precipitation correction applied as is usual. The X/Cl ratios for each of the precipitation sources were assumed to be constant throughout the year. For Ca, the average percentage of the total annual dissolved flux derived from precipitation was 10% with a range of 2% (January) to 25% (end of summer). Since the three sites were sampled at different times of day, and a diurnal melt cycle exists, an additional correction should also be applied to take into account the diurnal change in X/Cl ratios in order that data from the different sites are directly comparable. However, the data could not be corrected to take into account the time of sampling due to incomplete discharge data from Sites E and B.

A second precipitation input correction was calculated by using the meteorology station data, which recorded the volume of precipitation reaching the forefield (Fig. 2.5). The precipitation volume over two weeks was multiplied by the composition of the relevant rain or snow sample (ice melt was not included in this calculation). These were then added up to give an annual input flux from precipitation, which was subtracted from annual discharge fluxes obtained from the raw data and compared with annual fluxes obtained from the chloride corrected data (Table 3.4). These two values show very good

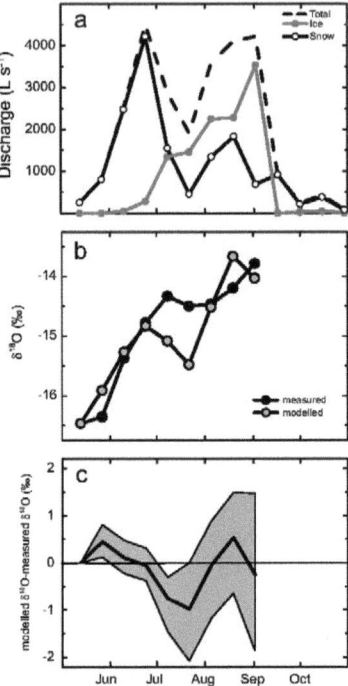

Figure 3.2: (a) Proportion of snow melt and ice melt contributing to discharge on the different sampling days as modelled by ALPINE3D [Magnusson et al., 2011]. (b) Comparison of modelled $\delta^{18}O$ compared to the measured values. The model is based on Rayleigh fractionation of remaining snow as the snow pack melts mixing with ice of a constant isotopic composition as described in the text. (c) The difference between modelled and measured $\delta^{18}O$. The grey area indicates the combined uncertainty of 20% in the estimation of snow and ice proportions, 10% in the estimation of catchment snow cover and 1‰ in the $\delta^{18}O$ value of ice.

agreement with each other, indicating that the precipitation correction is robust. The correction is quite small because dilute snow is the dominant precipitation source to this catchment. The precipitation correction decreased absolute concentrations (Fig. 3.1) but did not affect overall trends in chemical ratios (Fig. 3.3).

3.3.4 Stream water chemistry

The cation abundances were typical for glacial meltwaters draining alpine glaciers [Anderson et al., 1997a], with $Ca^{2+}>K^+\approx Na^+>Mg^{2+}$. Anion abundances were $HCO_3^->NO_3^->SO_4^{2-}\approx F^-\approx Cl^-$ (Table 3.1). Although the stream waters were dilute, significant spatial, seasonal and diurnal trends were observed in both precipitation corrected and uncorrected data (Figs. 3.1 and 3.3).

Although the three sampling sites lie within 1 km of each other and drain a common lithology, spatial differences were observed. The most pronounced spatial differences were observed in $^{87}Sr/^{86}Sr$ (Fig. 3.4): Site B was the most radiogenic with an average $^{87}Sr/^{86}Sr$ of 0.73024 and Site E was the least radiogenic with an average $^{87}Sr/^{86}Sr$ of 0.71838. The side stream (Site B) had more dilute major element concentrations than the main stream (Sites A and E) for the majority of samples (Fig. 3.1, Table 3.1). Spatial variation was also observed in element ratios (Fig. 3.3). Due to diurnal variability, part of the observed differences between sites could have been caused by sampling the sites at different times of the day (Table 3.1).

The concentrations of the major elements exhibited marked seasonal and diurnal variations in response to changes in discharge (illustrated by Ca^{2+} and F^- in Fig. 3.1). In general, the diurnal variation of the analysed parameters was $\sim 20\%$ of the seasonal variation. Maximum and minimum seasonal concentrations were observed in winter and at the end of August respectively. Over a single day, maximum concentrations were observed at night and minimum concentrations were observed during mid-afternoon. Discharge varied by a factor of over 200 over the year, whereas the maximum elemental concentration variation, observed for Si, was only a factor of 10. A similar muted response of concentration variations to discharge variations was observed at the diurnal timescale and demonstrates that the variability in solute concentrations is not only controlled by dilution. The attenuated response of solutes to changes in discharge has been termed 'chemostatic' in the hydrological literature [Godsey et al., 2009, Clow and Mast, 2010]. However, seasonal variations were observed in element ratios such as Ca/Si (Fig. 3.3) and such variations would not be expected if this catchment were behaving chemostatically.

At all three sites, Ca/Si ratios varied by a factor of 3-5 over the season (Fig. 3.3), with the lowest values observed in winter and the highest values observed during May/June when snow melt occurred. The seasonal change in Ca/Si ratios has also been observed in other glacial catchments [Hosein et al., 2004, Tipper et al., 2006a, Gabet et al., 2010]. Ca/Si also varied over a diurnal timescale, with the highest values coinciding with maximum discharge. The diurnal variation was most pronounced in August when the diurnal discharge amplitude was largest (2100 L s^{-1} compared to 300 L s^{-1} in June). The high concentrations of Si compared to Ca (low Ca/Si$_{SiteA}$ ratios of 0.45 - 1.53) are unusual with respect to previously published data on glacial catchments [e.g. Anderson et al., 2000]. Elevated Ca/Si ratios e.g. Ca/Si = 12 (Anderson et al. [2000], Bench River) are thought to be caused by calcite dissolution. The low streamwater Ca/Si ratios, together with the absence of calcite in this catchment [de Souza et al., 2010], strongly suggests that this catchment is not affected by calcite dissolution. Calcite precipitation is unlikely to occur as stream waters were always undersaturated with respect to calcite (SI$_{calcite}$ <- 4.4). Thus, the seasonal variation in Ca/Si ratios is unlikely to be caused by the changing proportion of carbonate to silicate weathering as proposed by Tipper et al. [2006a] for

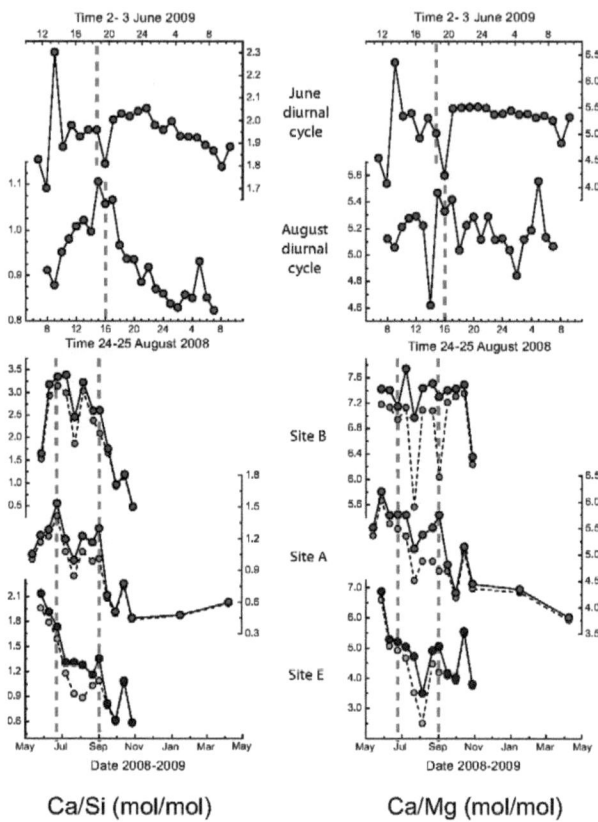

Figure 3.3: Seasonal and diurnal variations in two element ratios: Ca/Si (left column) which exhibits strong temporal variability and Ca/Mg (right column) which exhibits weak temporal variability. Red symbols are data from Site A, green symbols are for Site B and blue symbols are for Site E. Precipitation corrected values are indicated by orange, light green and light blue for Sites A, B and E respectively and are joined by a dashed line. The vertical dashed lines highlight the times of maximum discharge. Average Ca/Si$_{rock}$ was 0.06 (corrected for quartz content which is assumed not to weather) and average Ca/Mg$_{rock}$ was 1.3 [de Souza et al., 2010]

Himalayan rivers.

Although a subset of element ratios exhibit similar temporal trends to Ca/Si, other element ratios exhibit different temporal responses. On a diurnal timescale Ca/Mg shows essentially invariant behaviour (Fig. 3.3). On a seasonal timescale there is no variation in Ca/Mg at Site B but at Site A similar variation to Ca/Si is observed. The complex and varied temporal responses of element ratios are due to different sources (e.g. preferential mineral dissolution) and processes (e.g. ion exchange, formation of secondary phases) which act on each element. In addition, element ratio variations observed at Site A could also be affected by differing relative inputs from Sites B and E.

Significant temporal variation was also observed in strontium isotopes (Fig. 3.4). At all three sites there was a seasonal variation in $^{87}Sr/^{86}Sr$ of 0.006, with an increase from unradiogenic values at the start of June to more radiogenic values at the end of August. After 2nd September $^{87}Sr/^{86}Sr$ at Site B decreased whereas at Site E it continued to increase. Site A reflects values which are a mixture of the isotopic fluxes at Sites B and E (discussed further in Section 3.4.2.1). Diurnal variation in $^{87}Sr/^{86}Sr$ in August was 0.001 with an increase in values from mid-morning to mid-afternoon. The temporal trends in $^{87}Sr/^{86}Sr$ were not related to variations in discharge or element ratios but they do appear related to the temporal trends observed for $\delta^{18}O$ (compare Figs 3.1 and 3.4).

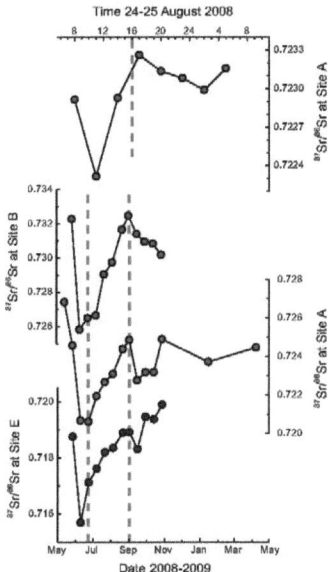

Figure 3.4: Seasonal and diurnal variation in $^{87}Sr/^{86}Sr$. The dashed lines highlight the times of maximum discharge. $^{87}Sr/^{86}Sr$ decreases in response to the snow melt peak in discharge at the start of the season but does not decrease in response to high discharge later in the year. Similar variation is observed on the diurnal timescale but with a smaller magnitude. The temporal trends in $^{87}Sr/^{86}Sr$ are similar to those observed in $\delta^{18}O$ (Fig. 3.1).

3.4 Discussion: water sources, solute sources and weathering processes

In the following discussion we will refer to water sources and chemical sources. We define *water sources* based on hydrology and $\delta^{18}O$, and *chemical sources* based on major element chemistry and $^{87}Sr/^{86}Sr$. The chemical composition of each water source can be measured at its origin, but along its flowpath its chemical composition can change and it can mix with other water sources, complicating source apportionment.

3.4.1 Identifying water sources

Identifying the sources of water and their flowpaths is crucial to understanding how rivers acquire their solutes [Malard et al., 1999, Brown et al., 2006], but is often neglected in many studies of chemical weathering. The ALPINE3D hydrological model [Magnusson et al., 2011] showed that snow and ice melt were the principal water components of this catchment, with the first half of the summer dominated by snow melt and the second half dominated by ice melt (Fig. 3.2a). Evapotranspiration and sub-surface components are minor contributors to the water budget of this catchment and can thus be neglected. Evapotranspiration contributed less than 3% of the water budget for this catchment [Kormann, 2009] and an additional sub-surface component was not required to accurately model the annual hydrograph. Variation in $\delta^{18}O$ reflects the changing hydrological properties of the catchment and the increase in $\delta^{18}O$ over the season, observed at all three sites, is typical for snow covered catchments [e.g. Bottomley et al., 1986, Unnikrishna et al., 2002, Welp et al., 2005]. The $\delta^{18}O$ values of the water inputs to the catchment can be used to provide information on the temporal changes of the water source contributions to the stream and aid in the interpretation of the observed chemical changes.

The validity of the source apportionment derived from the ALPINE3D hydrological model was tested by using it to predict the $\delta^{18}O$ value of the river water, and comparing these values to the observed values. Since the response of the catchment to rain events is very rapid (cf Fig. 2.5) and it did not rain on any of the sampling days, rain was neglected as a water source, reducing the problem to snow and ice inputs. A snow pack is initially composed of isotopically distinct layers reflecting different precipitation events, but over time isotopic redistribution processes serve to vertically homogenise snowpack $\delta^{18}O$ values [Raben and Theakstone, 1998, Taylor et al., 2001, Unnikrishna et al., 2002]. These processes occur during snow crystal metamorphism as a result of melting, freezing and vapour transport of water within the snowpack. The observation that meltwaters are isotopically lighter than the snowpack is understood to occur as a result of the fractionation of oxygen isotopes during melting, with ^{18}O preferentially retained in the ice phase [Taylor et al., 2001, Unnikrishna et al., 2002].

To estimate the $\delta^{18}O$ composition of water derived from the partial melting of snow,

Discussion: water sources, solute sources and weathering processes 37

it is necessary to take into account the isotopic fractionation between the solid phase (snow/ice) and water. This process can be represented by a Rayleigh fractionation process, where the $\delta^{18}O$ of the meltwater at time t ($\delta^{18}O_t$) is a function of the fraction of snow remaining (f).

$$\delta^{18}O_t = (\delta^{18}O_0 + 1000)f^{\alpha_{water-ice}-1} - 1000 \tag{3.1}$$

The equilibrium fractionation factor between water and ice ($\alpha_{water-ice}$) was taken to be 0.9965 at 0°C [Gat, 1996]. The fraction of snow remaining in the catchment (f) was modelled by Farinotti et al. [2010] based on daily photographs of the catchment. It is assumed that on 13th May, f is 1 and thus $\delta^{18}O_0$ is taken to be -16.47‰ (Table 3.1). Variations in $\delta^{18}O$ can occur with altitude, with lower $\delta^{18}O$ values at higher elevations [Gat, 1996], but as no systematic variations with altitude were observed, we assumed a homogeneous snow-pack.

The isotopic composition of snow melt will be further modified by mixing with ice melt. Since the fraction of ice removed compared to the total glacial ice volume is negligible, a constant $\delta^{18}O$ for ice of -16.66‰, based on seven melted ice samples from the dead ice block, was assumed. The estimated $\delta^{18}O$ values of snow melt and ice melt were combined according to the proportions derived from the ALPINE3D model [Magnusson et al., 2011] to estimate the bulk $\delta^{18}O$ of the river water.

$$\delta^{18}O_{river} = \delta^{18}O_{ice} \cdot F_{ice} + \delta^{18}O_{snow} \cdot F_{snow} \tag{3.2}$$

where F is the fractional contribution to discharge and $F_{ice} + F_{snow} = 1$.

The modelled seasonal variation of $\delta^{18}O$ in the river was compared to the measured values up to 2nd September. The predicted and observed $\delta^{18}O$ values agree remarkably well for such a simple model, with less than 1‰ discrepancy (Fig. 3.2b and c). This implies that the systematic increase in $\delta^{18}O$ over the season can be adequately explained by inputs from snow melt, controlled by fractional melting, mixing with ice melt. Discrepancies with the observed data could arise for a number of reasons. Firstly, there was an estimated error of 20% in the proportions of snow to ice melt from the ALPINE3D model. Secondly, the $\delta^{18}O$ value of glacier ice was not well constrained. Thirdly, rain inputs were neglected. Although it did not rain on any of the sampling days, water from previous rain events could still be percolating through the forefield. Fourthly, the Rayleigh approximation assumes the snow pack is a well-mixed, uniform reservoir; a state which is unlikely to be maintained throughout the melt season. Fifthly, during the percolation of meltwater through the snowpack and underlying ice, further isotopic exchange can occur [Taylor et al., 2001, Lee et al., 2010].

Although the retreat of the snow pack is important for the balance of snow to ice melt, it is also important to consider glacial drainage (though this will not change $\delta^{18}O$) as this will determine the contact time between water and sub-glacial sediments, potentially affecting the chemical composition of the water. The form of the hydrograph can help to infer glacial drainage. With its lower albedo, ice melts much more rapidly than snow and accentuates the daily discharge amplitude [Fountain, 1996], whereas snow cover attenuates daily discharge amplitude. The rapid melting of ice contributes to the development of a channelised (short water residence time) sub-glacial drainage system which also serves to accentuate the daily discharge amplitude [Nienow et al., 1996]. Thus, the difference between the maximum and the minimum discharge (daily discharge amplitude) each day reflects the amount of snow cover on the glacier and the nature of the subglacial drainage system. In 2008 the daily discharge amplitude increased up to mid-September reflecting the retreat of the snow-line and thereafter abruptly decreased in response to new snow. At the start of the melt season the drainage capacity of the glacier is low due to contracted channels and the glacier cannot drain water at the same rate as it is supplied from the glacier. Drainage capacity increases as the season progresses as melt water melts the ice causing channel expansion [Schuler et al., 2004]. This change in drainage can be inferred from the time difference between maximum solar insolation and maximum discharge [Fountain, 1996] and this time difference reached its minimum value at the end of August [Kormann, 2009], implying that this was when the subglacial network was draining most efficiently. These two parameters together point to an expanding channel network and decreased snow cover up until the beginning of September. Thereafter, the combination of a reducing channel network and new snow fall in mid-September caused the drainage efficiency and resulting discharge (Fig. 2.5) to sharply decrease. Thus, although $\delta^{18}O$ did not exhibit large variation during September (the sources remained the same), the way the water drained had changed. This sudden 'shutdown' of the glacial drainage system induced noticeable changes in the streamwater chemistry, as discussed below.

3.4.2 Identifying chemical sources of solutes

Whilst $\delta^{18}O$ and the ALPINE3D model show that the main water sources in this catchment are snow and ice melt, there could be additional water sources such as porewater and groundwater which are concentrated in solutes but are negligible for the total water balance. Even with only two water sources, variation in flow path length could create a number of different chemical sources through changes in the elemental composition as a result of processes such as dissolution, secondary mineral precipitation, exchange and biological cycling.

Discussion: water sources, solute sources and weathering processes 39

3.4.2.1 Sources of $^{87}Sr/^{86}Sr$

The $^{87}Sr/^{86}Sr$ ratio of the stream reflects the mixing of sources with different $^{87}Sr/^{86}Sr$ ratios, for example, different minerals or external dust inputs. However, $^{87}Sr/^{86}Sr$ should be insensitive to secondary processes such as the precipitation of secondary phases. The $\delta^{18}O$ ratios of groundwater samples are similar to those of the river (Table 3.3) suggesting connectivity between the stream and shallow groundwater [Malard et al., 1999, Magnusson et al., submitted], which could influence $^{87}Sr/^{86}Sr$. If there are no Sr inputs from the forefield areas and Sr is conservative over short timescales then Site A should fall on a mixing line between Sites E and B in $^{87}Sr/^{86}Sr$ vs 1/[Sr] space for each sample [Langmuir et al., 1978]. The minimum deviation between the mixing line and Site A data points can be calculated as the perpendicular distance from the mixing line. The calculated deviation from conservative mixing is small (<1 fmol L^{-1} Sr and an average $^{87}Sr/^{86}Sr$ deviation of 0.00135‰) and shows no seasonal trend (not shown). Much of the difference can be attributed to difference in sample collection times: $^{87}Sr/^{86}Sr$ at Site A had a diurnal variation of $\sim$0.001.

Ground and porewaters could contribute to the water mixture at Site A. However, porewaters have extremely heterogeneous chemical compositions (Table 3.3) and do not plot on the mixing line between sites E and B in Fig. 3.5. The $\delta^{18}O$ of the porewaters varied between rain values and river values suggesting that some of the porewaters were isolated with no connectivity to the stream. The groundwater samples were neither chemically nor isotopically distinct from the stream water samples (Fig. 3.5) and, as the volume contribution to total discharge was minor (Section 3.4.1), will consequently have an imperceptible effect on the stream water chemistry observed at Site A. Although we cannot completely exclude a contribution from ground and pore waters, it is likely that the forefield soils between Sites E and A have a negligible impact on the dissolved flux of Sr and this should be true for other elements which have similar chemical behaviour to Sr. Thus, for Sr, Site A can be explained as a simple mixture of waters from Sites E and B.

Previous measurements of $^{87}Sr/^{86}Sr$ in mineral separates and rocks from the catchment indicated a large degree of heterogeneity [de Souza et al., 2010]. It is thus likely that differences in the degree of metamorphic resetting of radiogenic Sr are controlling the spatial difference in $^{87}Sr/^{86}Sr$ between Sites E and B rather than different weathering conditions in each of the sub-catchments. For example, in terms of $^{87}Sr/^{86}Sr$ ratios (Fig. 3.5), Site E appears to drain a lithology more akin to the gneissic rock sample R04 (0.71613) and the mylonitic rock sample R05 (0.75577) is more representative of Site B [Fig. 3.5, de Souza et al., 2010]. Due to the large heterogeneity of $^{87}Sr/^{86}Sr$ in the mineral separates measured by de Souza et al. [2010] we cannot ascribe the stream water chemistry to a specific composition of minerals weathering.

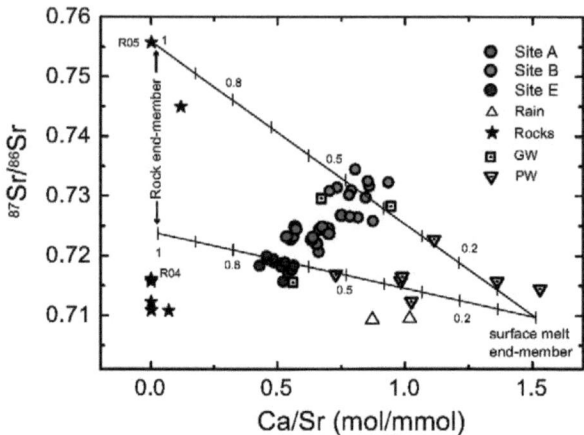

Figure 3.5: Mixing plot of $^{87}Sr/^{86}Sr$ against [Ca]/[Sr]. Site E can be explained as a mixture between average rock and a surface melt component. The surface melt end-member has a Ca/Sr ratio which is an average of snow samples and a $^{87}Sr/^{86}Sr$ ratio which is the average of the rain samples (Table 3.1). We assume $^{87}Sr/^{86}Sr_{snow} = {}^{87}Sr/^{86}Sr_{rain}$. Similarly, Site B can be explained by mixing of the surface melt component with the most radiogenic rock sampled (R05). Site A is a mixture between Sites B and E. The numbers on the mixing lines refer to the fractional contribution of Sr from rock. Rock data are from de Souza et al. [2010]. Porewaters (PW) have distinct compositions and have negligible influence on the stream water composition at Site A. Groundwater samples (GW) have identical compositions to the stream water indicating connectivity between the two [Magnusson et al., submitted], though they will not influence the stream water composition at Site A.

Significant dust deposition, principally from Saharan dust storms [De Angelis and Gaudichet, 1991, Thevenon et al., 2009], occurs along the Alpine chain. Saharan dust contains 20-50% carbonate [Goudie and Middleton, 2001] which is dissolved during transport by atmospheric aerosols [De Angelis and Gaudichet, 1991]. The addition of dissolved carbonates could impact stream water chemistry at Damma. However, the comparatively low Ca/Si ratios observed suggest that carbonate dust deposition does not strongly influence the stream water chemistry of this catchment.

Seasonal trends in $^{87}Sr/^{86}Sr$ have been observed in other catchments [Aubert et al., 2002, Krishnaswami et al., 1992, Tipper et al., 2006a, Bélanger and Holmden, 2010], and these have been linked to discharge control of weathering sources. It was proposed that discharge controls which chemically distinct areas of a catchment [Aubert et al., 2002] and which minerals [Bullen et al., 1996, Tipper et al., 2006a] contribute to the dissolved load. In this fieldsite, the lowest flows have the most radiogenic $^{87}Sr/^{86}Sr$ ratios which is broadly consistent with previous studies [Bullen et al., 1996, Tipper et al., 2006a]. However, a simple discharge control cannot explain the $^{87}Sr/^{86}Sr$ data since discharge is not correlated with $^{87}Sr/^{86}Sr$ ($R^2{}_{SiteA} = 0.17$).

Discussion: water sources, solute sources and weathering processes

The seasonal variation in $^{87}Sr/^{86}Sr$ at Damma is identical in form to that which would be predicted by the rapid addition and subsequent depletion of an unradiogenic source [Nezat et al., 2010]. We demonstrated in the previous section that $\delta^{18}O$ can be explained by the relative degree of snow and ice melt from the forefield and the glacier surface. If the 27th May sample is excluded then there is a linear relationship between $\delta^{18}O$ and $^{87}Sr/^{86}Sr$ for all sites (e.g. Site E, $R^2 = 0.62$). The correlation of $^{87}Sr/^{86}Sr$ with $\delta^{18}O$ suggests that the variations in $^{87}Sr/^{86}Sr$ can similarly be explained by snow melt. Total discharge is composed of snow melt and ice melt. We can consider that a fraction of this flow (f^*) comes into contact with sub-glacial sediment where it acquires solutes and $^{87}Sr/^{86}Sr$ ratios derived from dissolution, hereafter termed *sub-glacial* flow. Although two main types of subglacial drainage system exist [channelised and distributed, Nienow et al., 1996], we assume that the isotopic composition of strontium released from weathered sediments is unaffected by the nature of the sub-glacial drainage system. The rest of the discharge is en-glacial (within the glacier) or supra-glacial (on the glacier surface) flow where there is negligible solute acquisition. We term these last two components *surface melt*. As explained above, we assume that the groundwater and porewater inputs are negligible. The following mass balance equation can be written for Sr isotopes:

$$\phi_t[Sr]_t R_t = f^*(\phi_{ice}[Sr]^*_{ice} R^*_{ice} + \phi_{snow}[Sr]^*_{snow} R^*_{snow})$$
$$+ (1-f^*)(\phi_{ice}[Sr]_{ice} R_{ice} + \phi_{snow}[Sr]_{snow} R_{snow}) \quad (3.3)$$

where ϕ is the discharge, the subscript t is for 'total' and the asterisk indicates concentrations ([Sr]) and isotope ratios (R) modified by interaction with sub-glacial sediments. We assume ice has the same isotopic composition and Sr concentration as snow and the combined term is given the subscript g for 'glacial'.

$$\phi_t[Sr]_t R_t = f^* \phi_t[Sr]^*_g R^*_g + (1-f^*)\phi_t[Sr]_g R_g \quad (3.4)$$

If the concentrations and isotopic compositions of the two mixing components are assumed to be constant throughout the season then the value of f^* can be determined by fitting equation 3.4 with the values of R_t, ϕ_t and $[Sr]_t$ observed at Site A. The physical plausibility of the calculated values of f^* can then be assessed. The values of $[Sr]_g$ and R_g are assumed to be constant with values of 2.35 nmol L^{-1} (average snow [Sr], Table 3.1) and 0.70971 (snow $^{87}Sr/^{86}Sr$, from de Souza et al. [2010]) respectively. The choice of these values is supported by the fact that stream water data for Sites B and E lie on mixing lines between the surface melt end-member and rock compositions in $^{87}Sr/^{86}Sr$ vs Ca/Sr space (Fig. 3.5). During winter, the contribution of surface melt is expected to be negligible and these winter samples should represent the sub-glacial component. Thus, 74 nmol L^{-1} for $[Sr]^*_g$ and 0.73248 for R^*_g were chosen as these were the highest values observed in the stream water during winter. The calculated amounts of sub-glacial and surface meltwater compared to the total discharge are illustrated in Fig. 3.6.

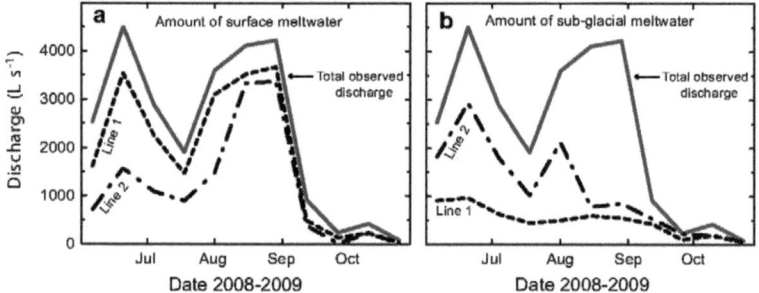

Figure 3.6: The calculated amount of surface meltwater (a) and the amount of sub-glacial meltwater (b) with respect to the observed total discharge. In both figures line 1 (dotted) represents values obtained by fitting equation 3.4 to observed values, assuming fixed chemical and isotopic compositions of the two components (see text). In (a) the values of line 2 (dash-dot) are the difference between the minimum diurnal discharge and the discharge at the time of sampling, assumed to represent surface glacial melt only. The values of line 2 in (b) are the minimum discharge values observed on the sampling day, assumed to represent sub-glacial meltwater only. The discrepancy between the two sets of values at the start of the melt season is likely to be due to snow-melt causing elevated diurnal discharge minima, leading to an overestimation of the amount of water which has interacted with sub-glacial sediments.

An independent approximation of the proportion of sub-glacial water can be obtained from the shape of the diurnal hydrograph. Diurnal discharge variations are driven by day-time melting of the glacier surface. It is assumed that the difference between the discharge at the time of sampling and the night-time minimum represents addition of surface melt which does not weather sediments ($1-f^*$). Although some of this water will inevitably reach the bed of the glacier, its effect, on a diurnal timescale, is to enlarge existing channels [Schuler et al., 2004], thus increased water-sediment contact will be minimal. Minimal melting of the glacier surface is expected to occur at night, therefore we assume that the diurnal minimum represents sub-glacial flow only (f^*). These values of f^* agree with the fitted values in the latter half of the season but underestimate the proportion of surface melt water at the start of the melt season (Fig. 3.6). During this part of the year, snow melt from the glacier surface dominates and maintains high flow rates even at night, due to the long time lag (>7 h) between maximum runoff and maximum solar insolation. Snow melt at night leads to an overestimation of the amount of sub-glacial water at this time of the year.

Although further data would be needed to verify the assumptions made, to a first approximation, the fitted amounts of surface and sub-glacial meltwater are plausible based on current understanding of glacier hydrology and the strontium isotopic composition of the stream water can be explained by a two-component mixture of surface melt and

Discussion: water sources, solute sources and weathering processes 43

subglacial water, whose composition is controlled by local lithology (Fig. 3.5). The relative proportions of these two solute sources are controlled by the evolution of the glacial drainage system over the melt season.

3.4.2.2 Elemental sources

Elemental ratios did not correlate with Sr isotopes, which only trace chemical sources. This implies that processes (e.g. adsorption) play an important role in modulating stream water chemistry in addition to chemical sources. The diurnal changes in stream water chemistry are very similar to those observed seasonally. We thus infer that the underlying controls on stream water chemistry are the same at both the diurnal and the seasonal timescales.

There are at least three processes which could alter elemental ratios within a water body: biological cycling, weathering and ion exchange.

Biological cycling could affect elemental ratios on diurnal and seasonal timescales in response to changing nutrient requirements. However the effect of biology on stream water chemistry in this catchment is expected to be negligible due to the sparse vegetation cover and the limited influence of porewaters on stream water chemistry (section 3.4.2.1). Further, stable isotope analyses of Sr and Ca have shown that the effect of biological cycling on river Sr and Ca is negligible [de Souza et al., 2010, Hindshaw et al., 2011a]. Weathering processes can affect element ratios by altering the ratio of ions released into solution during dissolution or through precipitation processes selectively removing ions from solution. Silicate dissolution rates vary with temperature and pH and the stream water temperature exhibited clear diurnal variation (Table 3.2). White et al. [1999b] reported the effect of temperature on individual elemental release rates from laboratory experiments. Using the temperature-elemental release rate relationships for Ca and Si derived for the most compositionally similar granite to Damma (weathered Loch Vale granite), we can calculate that an 8°C variation in temperature would be required to explain the observed diurnal variation in Ca/Si ratios in August. This change in temperature is much larger than the 3°C change observed. There was no clear diurnal variation in streamwater pH, therefore the influence of pH on Ca/Si ratios is expected to be minimal. Thus, changing silicate dissolution rates are unlikely to cause diurnal variations of element ratios. Ion exchange and sorption processes have rapid kinetics [Stumm and Morgan, 1996] and are known to cause the release of base cations during flushing with dilute water i.e. snow melt [e.g. Clow and Mast, 2010]. The effect of flushing with dilute water on element ratios such as Ca/Na appears to be catchment specific [Malard et al., 1999, Tranter et al., 2002b, Clow and Mast, 2010]. Ion exchange is undoubtedly important in this catchment but this process was unable to be quantified further.

Rather than the chemical composition of a single water body changing, the seasonal and diurnal variations could be caused by mixing different water bodies which have different chemical compositions [Malard et al., 2000]. Previous work on glacial drainage systems using dye tracers has identified two major components namely a 'distributed' component which is characterised by longer residence times and a 'channelised' component which is characterised by shorter residence times [Nienow et al., 1996, Brown, 2002]. The drainage system of the Damma glacier presumably behaves in a similar way.

Waters with different residence times have the potential to acquire different solute compositions. Laboratory experiments have shown that the initial stages of weathering are incongruent, resulting in the preferential release of Ca, Mg, K and Sr relative to Si and, to a certain extent, Na [Acker and Bricker, 1992, Brantley et al., 1998, White et al., 1999b]. Under high discharge conditions (short residence time), the stream water chemistry is likely to reflect incongruent weathering processes, whereas water with a longer residence time is expected reflect more congruent weathering. This prediction is in agreement with other fieldsites where high cation/Si ratios are often observed in water with short residence times compared to water with long residence times [Brown et al., 2006, Tipper et al., 2006a, Gabet et al., 2010]. It is therefore plausible that the diurnal and seasonal variations are caused by the variable mixing of water with different residence times and chemical compositions. For elemental ratios equation 3.4 would become:

$$\phi_t X_t = af^{**}\phi_t X_g^{*ch} + (1-a)f^{**}\phi_t X_g^{*dis} + (1-f^{**})\phi_t X_g \tag{3.5}$$

where X_g^{*ch} and X_g^{*dis} are the chemical ratios of the channelised (short residence time, more incongruent weathering) and distributed (long residence time, more congruent weathering) components respectively, a is the fraction of sub-glacial flow in the channelised system and f^{**} is the mixing proportion. With our data we cannot quantitatively constrain the fractional contributions of the channelised and distributed components but the observed changes in elemental ratios qualitatively support the mixing of waters with different residence times.

On a diurnal timescale, high meltwater discharge during mid-afternoon increases the channelised component [Stone and Clarke, 1996], resulting in high Ca/Si ratios compared to the night (Fig. 3.3). The Ca/Si diurnal variation is more pronounced in August compared to June due to the larger diurnal discharge amplitude. Both Ca and Mg are preferentially released during incongruent weathering [Erel et al., 2004]. Hence the Ca/Mg ratio is not expected to be strongly affected by variations in residence time and remains constant on this timescale (Fig. 3.3).

Similarly, changes in hydrological flowpaths can explain the observed seasonal variations. The channelised component increases in proportion as the melt season progresses due to increased melt forcing the expansion and merging of smaller sub-glacial channels [Tranter et al., 1996]. Thus, the channelised component is expected to dominate during summer (Fig. 3.7b) causing high Ca/Si ratios (Fig. 3.3). In autumn, colder

Discussion: water sources, solute sources and weathering processes

conditions and new snow block the melting of ice and the distributed flowpaths become more prominent (Fig. 3.7c) as evinced by the drop in Ca/Si ratios after 2 September (Fig. 3.3). In spring/early summer, distributed flow is expected to dominate the sub-glacial hydrological drainage system, but significant snow melt (which contains negligible Si) from the glacier surface also occurs (Fig. 3.7a), resulting in high Ca/Si values during this period. Seasonal, but not diurnal, variability was observed in Ca/Mg ratios at Site A (Fig. 3.3). This could be due to freezing of the forefield area during winter. Freezing of water concentrates solutes and CO_2 [Hodgkins et al., 1998]. High solute concentrations could cause the formation of secondary phases [Tranter, 2003] and Ca and Mg are likely to be affected differently by these conditions [Stallard and Edmond, 1983].

Glacial rivers tend to have high cation/Si ratios compared to non-glacial rivers [Anderson et al., 1997a, Hodson et al., 2000], and this has been attributed to the production of small, reactive particles as a result of glacial grinding. However, seasonal variations in elemental ratios e.g. Ca/Si are not confined to small glaciated catchments but are also observed in other rivers where present-day glaciation is negligible [Cameron et al., 1995, Yang et al., 1996, Aubert et al., 2002, Millot et al., 2002, Hren et al., 2007, Gupta et al., 2011]. Our data indicate that these variations are triggered by changes in discharge, which are amplified in glacial catchments.

In this lithologically homogeneous catchment significant variation was observed in elemental ratios, which are typically taken to indicate the degree of mixing between carbonate and silicate end-members [e.g. Gaillardet et al., 1999]. If an increased Ca/Na ratio was taken to indicate increased carbonate weathering rather than increased incongruency in silicate weathering, then cation release, and thus CO_2 consumption, from silicate weathering would be underestimated. For example, a typical formula for calculating silicate derived cation concentrations in meq is [Galy and France-Lanord, 1999]:

$$\sum cat^*_{sil} = Na^* + K^* + (2Na^* \times (Ca/Na)_{rock}) + (2K^* \times (Mg/K)_{rock}) \qquad (3.6)$$

where the asterisk refers to precipitation corrected data. If we take the average rock molar ratios, $(Ca/Na)_{rock}$ and $(Mg/K)_{rock}$, to be 0.38 and 0.28 respectively [Hindshaw et al., 2011a], and use the precipitation corrected annual fluxes from Table 3.4, then the calculated silicate cationic flux would be 66 meq/m^2/yr. This value is 44% less than that calculated assuming no carbonate contribution (117 meq/m^2/yr, Table 3.4). If this degree of underestimation applies to other rivers then our findings could have implications for current global estimates of silicate weathering rates.

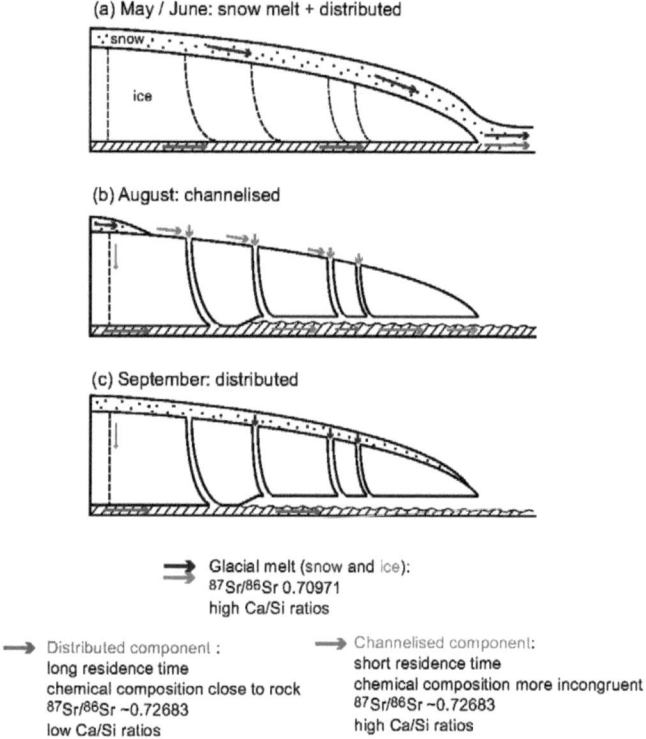

Figure 3.7: Conceptual model of how the glacier hydrology controls stream water chemistry represented by three different times during the season. Figure based on Brown [2002]. Blue arrows represent chemical contributions from snow and ice melt and red and green arrows represent chemical contributions from distributed and channelised sub-glacial flowpaths respectively. (a) At the beginning of the season snow still covered the glacier, but significant snow melt was occurring from the surface. This water mixed with concentrated water from the base of the glacier (distributed flow). (b) By August all snow in the ablation zone of the glacier had melted and crevasses had opened up which allowed icemelt to reach the bed of the glacier expanding sub-glacial channels and creating the fast channelised component which dominated the river chemistry. (c) In September, new snow fall which was not subject to strong melting due to cold temperatures, effectively sealed the crevasses (which gradually froze up) blocking transfer of melt water through crevasses. This resulted in a return to distributed flow conditions which persisted throughout the winter. During winter baseflow is maintained through pressure melting.

3.5 Conclusions

The stream water chemistry in this catchment is strongly controlled by glacial melt and the routing of this glacial melt water through the glacier, resulting in marked seasonal and diurnal variations in $\delta^{18}O$, $^{87}Sr/^{86}Sr$, element concentrations and element ratios. Hydrological modelling combined with oxygen isotope data showed that the two principal water sources in this catchment were snow and ice melt. The two primary chemical sources, as identified by strontium isotopes, were surface melt (consisting of supra-glacial and en-glacial snow and ice melt) and a mineral source derived from the weathering of sub-glacial sediments. This mineral source varied in its elemental composition depending on the residence time of the water in the glacial drainage system. Fast flow paths (channelised), which are dominant in summer and during the day, are characterised by high cation/Si ratios. Whereas slow flow paths (distributed), which are dominant in winter and at night, are characterised by low cation/Si ratios. The change in residence time could potentially influence several weathering related processes. Thus, the primary control on cation/Si ratios is the water flux and the pathway of that water through the glacier.

Our results emphasise that even in a notionally mono-lithological catchment, there is considerable spatial heterogeneity. The hydrologic control of temporal variability observed in this study is not confined to glacial catchments, suggesting that the primary impact of glaciers on stream water chemistry is through the modulation of discharge. The hydrological conditions at the time of sampling have to be taken into account when 1) comparing rivers sampled at different flow stages, 2) calculating annual chemical weathering fluxes based on a few spot samples and 3) deriving contributions of silicate weathering. Future stream sampling campaigns should consider sampling at various spatial and temporal resolutions in order to capture the sources of change and to further increase understanding of chemical weathering processes in riverine environments.

Chapter 4

A comparison of weathering fluxes from glaciated and non-glaciated granitic and basaltic terrains

4.1 Introduction

Large scale glaciation affects climate directly by altering the albedo of the land surface, and indirectly by changing sea levels and ocean circulation patterns [Imbrie et al., 1992, Sundquist and Visser, 2003]. Many feedbacks exist which determine the periodicity of glacial-interglacial cycles. One such negative feedback, which operates over long timescales, is thought to exist between climate and the chemical weathering of silicate rocks [Walker et al., 1981, Berner et al., 1983], whereby an increase in weathering could lead to reduced atmospheric CO_2 concentrations. Glaciers could potentially alter weathering rates compared to non-glaciated terrain, providing a pathway for weathering to impact atmospheric CO_2 levels [Gibbs and Kump, 1994, Sharp et al., 1995].

Modelling studies, which have focussed on the last glacial-interglacial transition, have reached conflicting conclusions regarding the impact of glaciers on global weathering rates. One model suggested that increased weathering at glacier margins would result in a decrease in atmospheric CO_2 [Gibbs and Kump, 1994], primarily as a result of the high runoff conditions at ice sheet margins. However, Ludwig et al. [1999] concluded that although total weathering rates would increase, those of silicates would not and thus there would be no effect on atmospheric CO_2. This result was corroborated by a study which included explicit treatment of glacier hydrology and chemistry [Tranter et al., 2002a]. Although chemical weathering rates may not change during glacial periods, transient increases in weathering rates could occur during de-glaciation due to increased runoff from ice melt and the exposure of reactive glacial debris [Vance et al., 2009]. Establishing the response of weathering to (de-)glaciation is important in order to understand the effect of glacial-interglacial cycles on chemical weathering rates.

Insight into past glacial processes can be gained from the study of currently glaciated terrain. It is assumed that weathering rates under large ice sheets [Gibbs and Kump, 1994] are negligible and this is confirmed by measurements from cold based glaciers which are frozen to their beds [Hodgkins et al., 1998]. Physical erosion and chemical weathering are intimately linked, with high physical erosion rates correlated with high chemical weathering rates [e.g. Riebe et al., 2004]. Glaciers have high physical erosion rates [Hallet et al., 1996] but whether they also cause high chemical weathering rates is not yet clear. Early studies of cation weathering fluxes from temperate glacial catchments suggested that glaciers enhanced weathering rates due to high, turbulent discharge, high physical erosion rates and highly reactive glacial flour which offset any reduction in weathering as a result of low temperatures [Reynolds and Johnson, 1972, Metcalf, 1986]. This view has been supported by more recent studies [Sharp et al., 1995, Oliva et al., 2003]. However, lithology is known to provide a strong control on the magnitude of weathering rates and often, the highly elevated cation fluxes were obtained from catchments where carbonates were known to be present [Sharp et al., 1995]. Other studies have found either no strong evidence for enhanced weathering from glaciated catchments [Anderson et al., 1997a, Hodson et al., 2000] or suppressed weathering rates [Gíslason et al., 1996, Hosein et al., 2004]. Anderson et al. [1997a] found that silicate weathering rates, as measured by Si fluxes, were reduced, but Hodson et al. [2000] argue that due to the heightened incongruency of weathering in glaciated environments, Si fluxes will underestimate true weathering rates.

The comparison of weathering rates between glaciated and non-glaciated catchments relies on the comparison of annual weathering fluxes. However, the calculation of accurate chemical fluxes from glaciated catchments is often logistically difficult. Stream systems are often braided and the channels shift from year to year making gauging challenging [Hodgkins et al., 2009]. Furthermore, corrections for precipitation inputs are more challenging due to the variable composition of snow melt and the mixing of snow melt with ice melt [Hodgkins et al., 2009, Hindshaw et al., 2011b]. In addition, it is often only feasible to obtain data from a short summer field season, requiring extrapolation to annual fluxes from limited data [Metcalf, 1986, Yde et al., 2005].

This study aims to assess the role of glaciers on weathering rates by first assessing the bias introduced by different methods of annual flux calculation using data from a small glaciated catchment in the Swiss Alps (Damma). Second, weathering fluxes and the chemical composition of stream water from lithologically identical glaciated and non-glaciated catchments are compared, in order to assess whether the weathering of these lithologies are affected by glaciers. The two lithologies investigated were basalt and granite.

4.2 Annual fluxes from the Damma catchment

Annual fluxes are used to compare catchments with each other, but are calculated by a variety of methods. To ensure that any observed inter-catchment differences are not due to differences in the method of flux calculation, the annual fluxes from the Damma glacier catchment were calculated using several different methods.

In this section, the results from different methods for calculating annual element fluxes are compared with each other using data obtained from the Damma glacier catchment [Hindshaw et al., 2011b]. The year used was 1 May 2008 to 30 April 2009 since this encompassed all the collected water samples. Discharge measurements were available from 6th June 2008 and discharge was recorded every ten minutes. The integrated discharge over the year was 2800±150 mm, with 90% occurring in a four month period (June to September). The discharge for May 2008 was estimated to be the average of April 2009 and June 2008.

Conductivity data was recorded at ten minute intervals from June to September 2008 and is often used as a proxy for total dissolved solids [Drever, 1997]. It can therefore be used to provide a high resolution concentration time series for each cation [Collins, 1983] and, in theory, more accurate fluxes. There were seven seasonal samples and 24 samples from a diurnal sampling campaign in August 2008 which had corresponding conductivity data. Conductivity and major cation concentrations were linearly correlated and the derived concentration-conductivity relationships (Fig. 4.1) were used to generate a concentration time series for each cation. The concentrations were multiplied by discharge to give a flux and the fluxes were integrated over time to give a total flux of each element over the period June to September. This flux was assumed to represent 80% of the total annual flux (see Method 3) and the scaled annual fluxes (Method 1) are reported in Table 4.1. The disadvantage of this method is that the conductivity-concentration relationships were based on comparatively few points and at high discharge or during heavy rainfall these relationships may be invalid. Nevertheless, it provides a method to capture overall daily and seasonal fluctuations in stream water chemistry [Metcalf, 1986].

Conductivity data may not always be available and the following methods are based only on the chemical data and discharge measurements. In Method 2 the annual flux was calculated by first calculating the instantaneous fluxes of the 15 seasonal samples collected and then the fluxes were integrated over time. The discharge for the two samples collected in May was estimated based on an interpolation between baseflow discharge and discharge values recorded in June. The same method was applied to precipitation corrected concentration data (Method 2*). The precipitation correction method is described in section 3.3.3. This flux calculation method cannot account for diurnal variability and is prone to bias due to sampling at similar times throughout the season. Nevertheless the calculated values (Method 2) are in good agreement with those

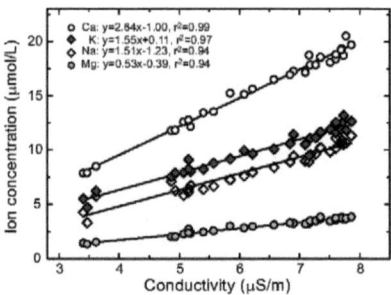

Figure 4.1: The relationship between ion concentrations and conductivity is linear and the derived relationships can be applied to the high resolution conductivity trace to obtain annual element fluxes.

calculated from Method 1 suggesting that any bias incurred from repeatedly sampling at high flow is minimal.

The following three methods (using precipitation corrected data) are all based on the assumption that concentrations are constant for a defined period of runoff. In Method 3, the monthly runoff is multiplied by the average monthly concentration (average of two to three samples) and the monthly fluxes are then summed. The average monthly concentrations for November, December, February and March are interpolated from those recorded in October, January and April. This calculation indicates that the period June to September accounts for 80% of annual solute fluxes. In Method 4, the concentration of each sample is multiplied by the runoff over the two week period centered on that date. This yielded a flux from June to October which was scaled to an annual flux assuming that the calculation period represented 80% of the annual solute flux. The final method applied (method 5), simply takes an average concentration multiplied by the total annual runoff. Methods 5a and 5b illustrate the range in flux values obtained if this method uses the minimum and the maximum concentrations respectively observed at Damma. As expected, method 5 yields higher fluxes than even the non-precipitation corrected data. This is because concentration and discharge are not linearly related and an average value based on few samples is unlikely to capture the full temporal variability of stream water concentration.

Apart from method 5, the precipitation corrected fluxes were within 15% of each other (and often less, depending on the element). The two non-precipitation corrected values were also within 15% of each other. This size of error can be expected given the $\sim$10% error in cation concentrations and the $\sim$10% error in discharge measurements. The large difference between fluxes calculated using annual discharge and average concentration data (Method 5*) and those calculated using finer resolution discharge and concentration data (Methods 2*-4*) suggests that finer resolution data are required to obtain accurate

Table 4.1: Summary of annual fluxes calculated by five different methods (described in text)

Method	Description	Ca	Mg	Na	K	Si	HCO$_3^-$	CDR†
		\multicolumn{6}{c}{kmol/km^2/yr}	meq/m^2/yr
1	Conductivity-concentration relationships	41	8	21	26	-	-	143
2	Integration of spot fluxes	36	7	21	24	34	64	131
2*		32	6	19	21	33	-	117
3*	Monthly averages	34	7	20	22	38	71	123
4*	Fortnightly averages	31	6	18	21	32	64	114
5*	Annual average	50	11	35	33	73	112	188
5a*	Annual min	16	3	7	10	16	37	57
5b*	Annual max	115	31	97	67	197	271	456
Average 2*, 3* and 4*		32	6	19	21	34	66	118

†Cation denudation rate (CDR) = 2*Ca + 2*Mg + Na + K

fluxes [Gupta et al., 2011]. With the present data set there are too few samples to assess the optimal time period with which to obtain the most accurate fluxes with the minimum number of samples. In addition, for long-term weathering rates, monitoring over a period of several years, preferably longer, is required in order to capture inter-annual variation in weathering fluxes. Nevertheless, if conductivity is used as a proxy then the relationship between conductivity and concentrations should be obtained with the maximum number of data points possible, covering a wide range of hydrological conditions. In the following discussion the average of methods 2*, 3* and 4* will be quoted. Many studies report fluxes based on method 5, and these are most biased towards the hydrochemical conditions at the time of sampling. However, the difference in calculated element fluxes between methods (less than a factor of two) is less than the difference in reported element fluxes between rivers draining a single lithology (two to three orders of magnitude), therefore reported fluxes based on method 5 can still be used in inter-river comparisons.

4.3 Comparison of glaciated and non-glaciated samples

Previous comparisons of annual weathering fluxes between glaciated and non-glaciated samples have not been strictly comparable in that lithologically similar catchments were not compared [Anderson et al., 1997a, Hodson et al., 2000]. In the following discussion we compare glaciated and non-glaciated catchments from the two lithologies on which most data exists: granite and basalt.

4.3.1 Sources of data

Four glaciated catchments have been included: two from the Swiss Alps - Damma glacier [Hindshaw et al., 2011b] and Rhone glacier [Hosein et al., 2004] and two from Greenland: Leverett glacier (Hindshaw, unpublished data) and Mittivakkat glacier [Hagedorn and Hasholt, 2004]. Data from non-glaciated granitic catchments is taken from the databases of Oliva et al. [2003] and White and Blum [1995]. Granite catchments where there is strong evidence for calcite dissolution or the presence of other outcrops have not been included. Catchments underlain by gneiss have been included and those underlain by diorite excluded.

Data for glaciated basaltic catchments is predominantly from Iceland and is taken from Gíslason et al. [1996] and the sample set collected by researchers from the Open University [OU group: Gannoun et al., 2006, Georg et al., 2007, Vigier et al., 2006]. In addition there is data from Kuannersuit Kuussuat, Greenland [Yde et al., 2005]. Not all the Icelandic catchments are glaciated and these together with data from the Deccan Traps [Dessert et al., 2001, Das et al., 2005, Jha et al., 2009, Gupta et al., 2011], Réunion [Louvat and Allègre, 1997], Azores [Louvat and Allègre, 1998] and the Paraná Traps [Benedetti et al., 1994] form the non-glaciated basaltic database.

Table 4.2 provides a brief description of the methods used by each study to determine precipitation corrections and annual element fluxes. Precipitation corrected element fluxes are reported in Table B.4.

The wealth of data from Iceland offers the opportunity to compare fluxes obtained by different workers from the same rivers. Although the concentration data from spot samples taken by the OU group [Gannoun et al., 2006, Georg et al., 2007, Vigier et al., 2006] were similar to those reported by Gíslason et al. [1996], some river fluxes were different because, due to inter-annual variability in runoff, different runoff values were used. For example the runoff of Hvítá-W at Kljáfoss is reported as 1685 mm by Gíslason et al. [1996] and 970 mm by the OU group. Similarly, Gupta et al. [2011] and Dessert et al. [2001] both report fluxes from the Narmada river which agree very well considering the different methods of flux calculation (Table B.4), probably because the majority of runoff occurs in the wet season. In addition, Gupta et al. [2011] show that fluxes computed from wet and dry season spot samples were comparable to average fluxes obtained from ten years' worth of fortnightly sampling.

4.3.2 The stream water chemical composition of glaciated and non-glaciated catchments

Due to the unique weathering conditions occurring in glaciated environments (high physical weathering, high discharge) it may be expected that the chemical composition of water emerging from the glacier has a unique chemical composition. Specifically, it has been proposed that weathering in glaciated environments is highly incongruent i.e. cations such as K in the interlayer sites of biotite are preferentially released [Anderson et al., 1997a, Blum and Erel, 1997]. Calcium concentrations have been found to be elevated in glaciated catchments and this is often attributed to the weathering of carbonate phases [Tranter et al., 2002b]. In this study we have not included sites where carbonate phases are known to be present, however we cannot rule out trace carbonate phases in all the catchments. To examine whether the stream water composition of glacial and non-glacial

Table 4.2: Summary of flux calculations and precipitation correction methods applied to a compilation of basaltic and granitic catchments

Locality	Source	Lithology	Flux calculation	Precipitation correction	Other comments
Iceland	Gislason et al. [1996]	B	Annual mean concentration calculated from 17 - 23 months of sampling multiplied by annual runoff.	Chloride correction using ratios derived from empirical relationships in local rain composition.	Mix of glaciated and non-glaciated catchments. Hot spring affected catchments excluded.
Iceland	Gannoun et al. [2006], Georg et al. [2007], Vigier et al. [2006]	B	Single spot sample multiplied by annual runoff.	Chloride correction following Gislason et al. [1996].	Mix of glaciated and non-glaciated catchments.
Kuannersuit Kuussuat (Greenland)	Yde et al. [2005]	B	A concentration - discharge relationship was derived from four days of sampling and applied to the rest of the ablation season.	Chloride correction using seawater ratios.	
Deccan Traps (India): Narmada and Tapi Rivers	Dessert et al. [2001]	B	Single spot sample multiplied by annual runoff, assumed identical runoff for all rivers sampled.	Chloride correction using seawater ratios.	Rivers draining Indian Shield excluded.
Deccan Traps (India): Narmada River	Gupta et al. [2011]	B	Monsoon and non-monsoon spot samples were multiplied by the average wet and dry season runoff respectively and added.	Chloride correction using measured precipitation.	
Deccan Traps (India): Godawari River	Jha et al. [2009]	B	Monsoon and non-monsoon spot samples were multiplied by 80% and 20% of the annual runoff respectively and added.	Chloride correction using measured precipitation.	Only upstream samples draining basalt included.
Deccan Traps (India): Bhima, Krishna rivers	Das et al. [2005]	B	Single spot sample multiplied by annual runoff.	Chloride correction using measured precipitation.	
Réunion	Louvat and Allègre [1997]	B	Single spot sample multiplied by annual runoff.	Chloride correction using seawater ratios.	Only rivers not affected by hot springs.
Azores	Louvat and Allègre [1998]	B	Single spot sample multiplied by annual runoff.	Chloride correction using average ratios from three local rain samples.	
Paraná Traps (Brazil)	Benedetti et al. [1994]	B	Wet and dry season spot samples were multiplied by the average mean wet and dry season discharge. Annual flux calculated assuming wet and dry season contribute six months each to total discharge.	Chloride correction using ratios from local rain.	

(Continued on next page)

Table 4.2 – Continued

Locality	Source	Lithology	Flux calculation	Precipitation correction	Other comments
Damma glacier (Switzerland)	Hindshaw et al. [2011b], this study	G	Detailed in section 4.2.	Chloride correction using variable ratios depending on source of precipitation (snow/ice/rain).	
Rhône glacier (Switzerland)	Hosein et al. [2004], Hosein [2002]	G	A flux was calculated for each sample and an average was calculated for each month. The monthly averages were summed to give an annual flux. A total of 191 samples were used in this calculation. Diurnal sampling ensured that both high and low flow discharge were sampled.	Chloride correction using composite ratios derived from 60% snow and 40% rain contribution.	
Leverett glacier (Greenland)	this study	G	A flux was calculated for each sample and 74 samples covering 28 days were integrated to give the flux for one month. Based on the hydrograph [Bartholomew et al., 2011] this period represented 25% of the annual discharge and the fluxes were scaled accordingly to give an annual flux.	No correction made.	
Mitivakkat Glacier (Greenland)	Hagedorn and Hasholt [2004]	G	Average concentration multiplied by estimated annual runoff based on 120 day ablation season.	Chloride correction using ratios from a snow sample.	Excludes spot samples draining Ammassalik Intrusive Complex.
Slave Province and Grenville Province (Canada)	[Millot et al., 2002]	G	Single spot sample multiplied by annual runoff.	Chloride correction using seawater ratios with adjusted chloride concentrations.	
Estibère, Escale, Trois Seigneurs (France)	Oliva et al. [2004]	G	Discharge weighted mean concentration multiplied by annual runoff.	Chloride correction using local precipitation ratios (Vielha) given in Camarero and Catalan [1996].	
Guayana	Edmond et al. [1995]	G	Discharge weighted average concentration multiplied by annual runoff.	Chloride correction using seawater ratios.	Only data from Parguaza.
Cameroon	Viers et al. [1997, 2000]	G	Single spot sample multiplied by annual runoff, assumed identical runoff for all rivers sampled.	Chloride correction using regional precipitation ratios given in Freydier et al. [2002].	Only rivers draining gneiss or granodiorite included.

(Continued on next page)

Comparison of glaciated and non-glaciated samples 57

Table 4.2 – Continued

Locality	Source	Lithology	Flux calculation	Precipitation correction	Other comments
Solmyren, Vuoddashbäcken, Lilla Tivsjön (Sweden)	Calles [1983]	G	Annual output flux measured from monthly stream samples minus the annual input flux from precipitation.	Annual precipitation input measured from monthly samples.	
Storbergsbäcken (Sweden)	Land and Öhlander [2000]	G	Annual stream output flux minus the annual input flux from precipitation. Calculation includes a term for nutrient uptake by vegetation.	Precipitation fluxes calculated from average of several rain and snow samples.	
Breidvikdalen (Norway)	Skartveit [1981]	G	Annual stream output flux calculated from weighted mean annual concentrations minus the annual input flux from precipitation.	Annual input flux derived from weighted mean annual concentrations.	
Schluchsee (Germany)	Feger et al. [1990]	G	Annual output flux measured from weekly stream samples minus the annual input flux from precipitation.	Annual input flux derived from weekly samples.	
Bärhalde (Germany)	Stahr et al. [1980]	G	Annual output flux measured from regular stream samples minus the annual input flux from precipitation.	Annual input flux derived from weekly samples.	
Margeride (France)	Négrel [1999]	G	Annual flux calculated from discharge and concentration measurements of three spot samples from different times of the year.	Chloride correction using mean of 15 monthly precipitation samples.	
Lysina (Czech Republic)	Krám et al. [1997]	G	Volume weighted stream water flux minus precipitation flux.	Volume weighted precipitation flux from bi-weekly samples.	
Tsukuba (Japan)	Hirata and Muraoka [1993]	G	Annual stream output flux calculated from 4 years worth of data minus precipitation flux.	Annual precipitation input flux calculated from 4 years worth of data (method and sampling frequency not given).	
Panther Lake (US)	April et al. [1986]	G	Annual stream output flux calculated from 3 years worth of data minus precipitation flux.	Annual precipitation input flux calculated from 3 years worth of data (method and sampling frequency not given).	
Loch Vale (US)	Mast et al. [1990]	G	Annual stream output flux calculated from long term weekly sampling of data minus precipitation flux.	Annual precipitation flux based on long term weekly sampling.	

streams were different, element ratios with Na as a denominator were plotted. Sodium was chosen as the denominator since it is the least likely of the four major cations to be preferentially released.

In basaltic catchments there was no difference between glaciated and non-glaciated catchments in Mg/Na vs Ca/Na space (Fig. 4.2a). Ratios from glaciated Icelandic catchments were identical to those from non-glaciated Icelandic catchments and those from the Azores. Similarly, glaciated and non-glaciated catchments from Iceland were indistinguishable in Mg/Na vs K/Na space (Fig. 4.2b).

The linear trend in Mg/Na vs Ca/Na ratios in rivers draining basalt has been attributed to differing basalt rock compositions rather than a silicate-carbonate trend caused by the weathering of calcite in basalt [Dessert et al., 2003]. This compilation supports this conclusion. Tholeiitic basalts are poor in Na compared to alkali basalts, thus rivers draining tholeiitic basalts (Deccan and Paraná traps) would be expected to have elevated Ca/Na and Mg/Na ratios compared to those draining alkali basalts (Azores). In Iceland the main rift produces tholeiitic basalts whereas the off-rift zone in the south-west, where the river samples of Gíslason et al. [1996] were collected, produces alkali basalts [Sigmarsson and Steinthórsson, 2007]. The basalts of Réunion are intermediate in composition between tholeiitic and alkaline and this is reflected in the Ca/Na and Mg/Na ratios which are intermediate between the Azores and the Deccan Traps.

In both basaltic and granitic catchments, for a given Mg/Na ratio, there is a large scatter of K/Na ratios, and several catchments had negative K concentrations after the correction for precipitation inputs. This is probably due to a combined effect of an incorrect precipitation correction, uptake by clay minerals and the influence of vegetation. The dissolved fluxes of K are strongly impacted by vegetation, resulting in considerable uncertainty in the determination of K weathering fluxes [Likens et al., 1994].

In granitic catchments, part of the linear trend in Mg/Na vs Ca/Na may be due to the influence of trace carbonate. However, the comparatively large spread of data from a single river (e.g. Rhone glacier) suggests that additional factors, other than lithology, are responsible for at least some of the variation observed. This could be hydrological control [Hindshaw et al., 2011b] or anthropogenic effects [Gupta et al., 2011]. The glaciated granitic catchments (excepting Mittivakkat glacier) had elevated Ca/Na and K/Na ratios for a given Mg/Na ratio compared to non-glaciated catchments (Fig. 4.2c, d). Enhanced carbonate dissolution would increase both Ca/Na and Mg/Na ratios, thus the difference between glaciated and non-glaciated catchments is more consistent with enhanced weathering incongruency in glacial environments [Hodson et al., 2000].

Elevated K concentrations in glaciated catchments are most likely due to the combined effect of enhanced biotite weathering caused by glacial abrasion [Anderson et al., 1997a] and the increase in the weathering rate ratio of biotite to plagioclase at lower temperatures [White et al., 1999b]. Biotite is only present in some basalts, and this is reflected in the

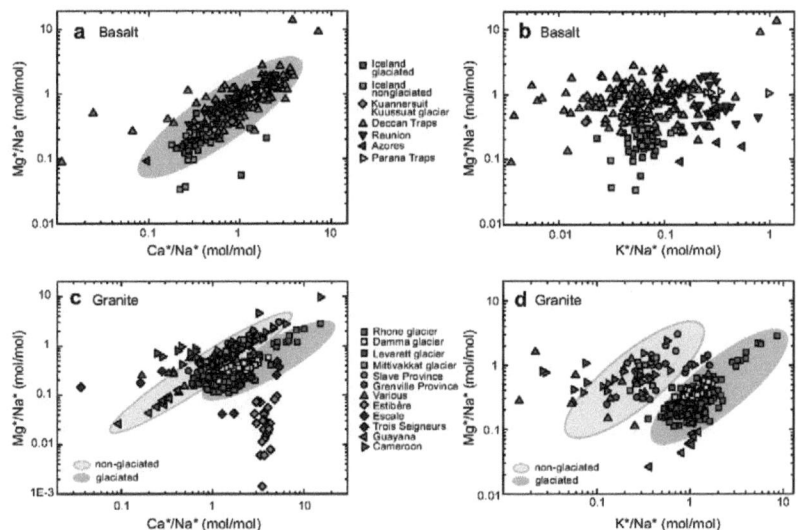

Figure 4.2: Element ratio plots for basaltic (**a** and **b**) and granitic (**c** and **d**) catchments. Data from Rhone, Leverett and Damma glaciers were seasonal spot samples and indicate the range of values obtained from a single river. The grey ellipses highlight the main trends in the data.

K/Na ratios which are a factor of 10 less than those observed in granitic catchments.

The low K/Na ratio of the Mittivakkat glacier, compared to the other glaciated catchments may be due to the geology of the region. Although the proglacial area drains granite/gneiss, this is surrounded by the Ammassalik Intrusive Complex (AIC) which is an enstatite-rich diorite. Due to the glacier obscuring the underlying geology, it is not clear what the relative contribution of each type of lithology is to the runoff composition. The AIC contains no potassium feldspar [Hagedorn and Hasholt, 2004], and this could be responsible for the comparatively low K concentrations, or the Mg-bearing enstatite could elevate Mg/Na ratios above those normally found in granitic catchments.

The Mg/Na ratios from the Estibère catchment were anomalous compared to the rest of the granitic catchments (Fig. 4.2c). This may be due to the precipitation correction. An average precipitation composition based on measurements from a location outwith the watershed (Vielha, Pyrenees) was used as the basis for the precipitation correction for Estibère, Escale and Trois Seigneurs [Oliva et al., 2004]. The latter two watersheds lie to the east of Vielha and Estibère to the west (further from the Mediterranean sea). The more inland location may be susceptible to precipitation with a different chemistry. In general, detailed precipitation measurements are lacking for several catchment studies. Assuming a sea salt composition may be valid for oceanic islands but location specific precipitation measurements are required for all other catchments and are just as important as the stream water samples themselves.

4.3.3 Are silicate weathering fluxes affected by glaciers?

Lithology strongly controls annual weathering fluxes, thus to compare the effect of glaciers on weathering rates, glaciated and non-glaciated catchments with the same lithology should be compared rather than comparing weathering rates to a continental mean.

The silicate weathering rates of glaciated basaltic catchments are comparable with non-glaciated catchments (Fig. 4.3a). Cation and silica fluxes are linearly related with glaciated and non-glaciated samples lying on the same trend suggesting that silicate weathering fluxes are not suppressed in glaciated catchments. In addition, temperature does not appear to be the main control of basalt weathering rates - the tropical Paraná Traps ($\sim 25°C$) have identical weathering rates (cation and silica fluxes) to Icelandic catchments ($\sim 5°C$). The majority of river samples from the Deccan Traps appear to form a separate trend with higher cation fluxes in comparison to other basaltic catchments with comparable silica fluxes. This could be because this was the only region where there is significant use of water for agriculture and irrigation and in which samples were collected below the location of dams. Long term data of annual fluxes before and after one of these dams (Sardar Sarovar dam) indicate that the dams can reduce silica fluxes by 45% and cation fluxes by 30% [Gupta et al., 2011]. The greater decrease for Si compared to cations could be responsible for the apparent offset of the Deccan rivers. Alternatively, the presence of saline-alkaline soils, outcrops of carbonate rocks and significant groundwater inputs [Gupta et al., 2011] may elevate cation fluxes. Four rivers from this region, which were sampled above any dams plot on the same trend as the other basaltic rivers indicating that dams could be the main cause of reduced silica fluxes. Two Icelandic catchments have elevated cation fluxes, but elevated sulphate concentrations suggest that these catchments have a hot spring influence.

As previously reported, absolute fluxes in granitic catchments are lower than those of basaltic catchments [Bluth and Kump, 1994]. Overall, cation and silica fluxes are correlated, but there are several catchments which lie off the main trend. The two glaciated catchments in Greenland have increased cation fluxes relative to the two Swiss glaciated catchments despite having similar Si fluxes. As discussed in section 4.3.5, the weathering of granite is sensitive to changes in the physical erosion regime. It is very likely that, due to the lower winter temperatures in Greenland, part of these glaciers are frozen to the bedrock [Huybrechts, 1996]. The most intense glacial erosion occurs in the zone next to cold-based ice, so-called warm-freezing ice [Bennett and Glasser, 1996]. The presence of this zone in Greenlandic glaciers and not in the warm-based Swiss glaciers could be responsible for the enhanced incongruency between cation and silica fluxes observed for

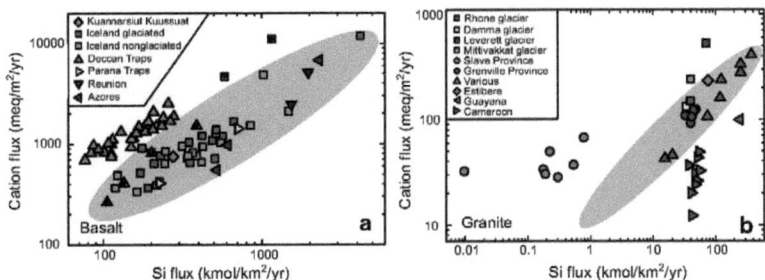

Figure 4.3: Annual cation flux vs annual Si flux for basaltic (a) and granitic (b) catchments. In (a) the Deccan Trap points marked by an internal white circle are those which are unaffected by upstream dams or pollution. The marked Icelandic samples are those with suspected hot spring inputs based on SO_4 concentrations. The grey ellipses highlight the overall trend in the data.

the Greenlandic rivers. Rivers draining the Slave province have exceptionally low Si fluxes, but comparable cation fluxes to Loch Vale and Storbergsbacken. Rather than representing low silicate weathering rates, the low Si fluxes are likely due to Si uptake by diatoms in the lakes [Engström et al., 2010, Opfergelt et al., 2011]. The Slave region is characterized by a large network of lakes, and rivers were sampled in the summer when Si uptake rates are likely to be high.

Rivers from Cameroon and Guayana drain shield terrain which was not glaciated during the last glacial maximum. The rest of the catchments in this compilation were all very likely glaciated at one point during recent geological history. The low rates of physical weathering and the transport-limited weathering regime could be potential reasons accounting for the different relationship between cation and silica weathering fluxes from these two catchments compared to the rest [Edmond and Huh, 1997]. Thus, for granitic catchments, it is not the extent of glaciation per se which controls the weathering fluxes but the erosional regime.

4.3.4 Climate

Although a plot of fluxes against runoff may be flawed [Gaillardet et al., 1999], it does highlight some differences between the rivers (Fig. 4.4). In the basaltic catchments, the Deccan rivers again form an outlier from the rest of the rivers. As discussed above this is likely due to the influence of dams in regulating the runoff, preventing the 'true' runoff from being measured. For granitic catchments, cation fluxes increase with runoff, but above ∼1000 mm runoff cation fluxes decrease with increasing runoff, this decrease is especially pronounced between the four glaciated catchments. This data is consistent with the idea that, above a certain flow-rate, there may be a change in the rate-limiting weathering process [Kump et al., 2000], as observed in a data compilation of world rivers [Holland, 1978]. This would mean that previous models of weathering based on a linear relationship between runoff and weathering rates [Gibbs and Kump, 1994] may need to be revised. It is unclear why a similar effect is not observed for glaciated basaltic catchments. One reason could be that, due to the higher intrinsic weathering rate of basalt, the transition in rate-limiting step will only be reached at higher runoff values.

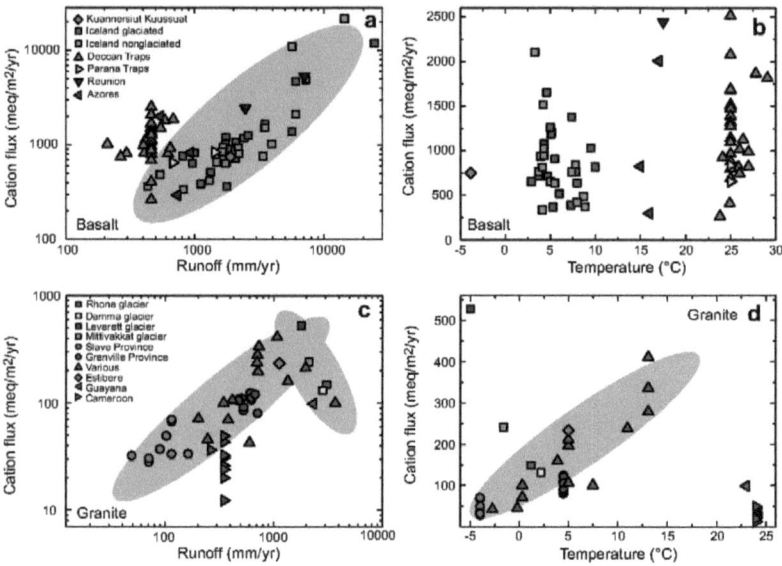

Figure 4.4: Annual cation fluxes vs runoff and mean annual air temperature for basaltic (**a, b**) and granitic (**c, d**) catchments. Similar plots are obtained if silica or bicarbonate fluxes are plotted instead of cation fluxes. The grey ellipses highlight the overall trends in the data.

In this compilation, no clear correlation of mean annual air temperature with weathering rates was found for basaltic catchments (Fig. 4.4b), in contrast to Dessert et al. [2001], which suggests that there are factors more important than temperature which determine inter-catchment variability in weathering fluxes. However, on a local scale, Gíslason et al. [2009] found statistically significant correlations between temperature and chemical weathering rates within individual Icelandic catchments over the past 44 years.

The shield terrain catchments, represented by Cameroon and Guayana, are thought to be insensitive to climate parameters because they are in a transport-limited weathering regime [West et al., 2005]. These catchments have slightly lower cation fluxes than weathering-limited catchments with similar runoff (Fig. 4.4c) and, based on the trend observed between cation fluxes and temperature within the weathering-limited catchments, much lower cation fluxes than would be expected given their temperature (Fig. 4.4d).

Excluding the two Greenlandic catchments, the remaining weathering-limited granitic catchments show a strong dependence of cation flux with temperature. The apparent activation energy (E_a') for silicate weathering can be calculated using the Arrhenius equation: the higher the value of E_a', the stronger the feedback between climate and weathering [Velbel, 1993]. This approach necessarily assumes that all other rate-affecting variables are identical between the catchments.

$$\ln Q_{Si} = \ln A - \frac{E_a'}{RT} \qquad (4.1)$$

In equation 4.1, Q_{Si} is the annual flux of Si, A is the pre-exponential factor and R is the gas constant. The apparent activation energy is obtained from the gradient in a plot of the natural logarithm of silica flux versus the reciprocal of absolute temperature (Fig. 4.5). This yields an E_a' of 117 ± 12 kJ mol^{-1}. This value is higher than previous estimates of E_a' from granitic watershed studies [60-77 kJ mol^{-1}, Velbel, 1993, White and Blum, 1995, Oliva et al., 2003, West et al., 2005] and is also at the high end of the range obtained during laboratory studies [Brantley, 2003].

The exact feedback between weathering and climate depends on the proton source and this may not be atmospheric CO_2 for weathering in a sub-glacial environment, where there is limited contact with the atmosphere [Tranter, 2003]. It has been suggested that microbial oxidation of overridden soil organic carbon and sulphides may provide the dominant proton source for weathering reactions in sub-glacial environments [Sharp et al., 1999]. The balance of proton supply between atmospheric CO_2 and microbially mediated oxidation, which is expected to be dependent on the size of the glacier, will determine the effect of glaciers on the feedback between weathering and climate and remains to be quantified.

Silicate weathering rates are also dependent on pH and most streams draining granitic catchments have a pH of 5-7. In contrast, the stream draining Leverett glacier had a pH of 8-9 (there was no pH data available for Mitivakkat glacier), caused by the lack of atmospheric equilibration under the ice sheet. The weathering rates of some silicate minerals (e.g. albite) have a parabolic dependency on pH with a minimum at pH 5-7 [Brady and Walther, 1989], thus the weathering reactions under Leverett glacier may proceed at a faster rate than in non-glaciated granitic catchments, assuming that the pH recorded at the glacier outlet reflects the pH of water under the ice. Streams draining basaltic catchments are neutral to alkaline (pH 7-8) and there is no discernable change in pH between the glaciated and non-glaciated Icelandic catchments. This could be because basaltic streams are more efficiently buffered than granitic streams due to faster dissolution rates [Gannoun et al., 2006].

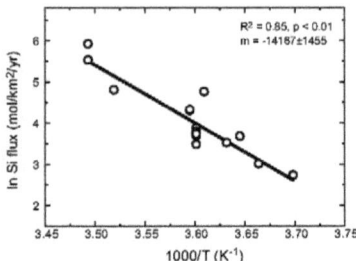

Figure 4.5: Plot of the natural logarithm of annual Si fluxes versus the reciprocal of mean annual temperature for the weathering limited granitic catchments, excluding Greenland (glacier effect) and Slave Province (Si likely affected by diatom uptake).

4.3.5 Lithology

Basaltic and granitic catchments clearly respond differently to glaciation. Basaltic catchments appear to be unaffected whereas glaciation changes the chemical composition and slightly elevates the cation fluxes of granitic catchments. Although both are silicate rocks they are fundamentally different in both their mineral composition and texture: granite is coarse-grained and massive whereas basalt is fine-grained, brittle and often contains gas bubbles. We suggest that the enhanced physical erosion during glaciation has little effect on basaltic catchments because the starting material is already finely crystalline and friable. On the other hand, because the starting material is coarse, physical weathering has a measurable impact on granite. The weathering rates of basalt are correspondingly faster due to the greater surface area. Thus, although both granite and basalt contain biotite, it is only in glaciated granitic catchments that K release from biotite is enhanced. A similar argument was invoked to explain the difference in weathering between granitic shield terrains and mountain belts. Stallard [1995] proposed that mountain belts weathered faster because the granite was less massive than granite in shield terrains. The frost shattering of rock through freeze-thaw cycles is another cold temperature process which would have similar physical effects to glaciation. Frost-shattering continually exposes fresh mineral surfaces and fractures the rock, altering element ratios [Hoch et al., 1999] and increasing weathering rates [Huh et al., 1998]. The lack of enhanced weathering rates in the arctic Mackenzie river basin [Millot et al., 2003] could be explained by the fact that a large part of the river system drains shale. Shale is fine-grained rock and, based on the above argument, would be expected to be less sensitive than granite to the effects of cold temperature physical erosion processes. The apparent close similarity in the weathering rates from glaciated and non-glaciated granitic catchments probably arises because the majority of these catchments are located in cold or mountainous regions which are presently, or were once, affected by either glaciation or frost-shattering. Subtle changes in the texture of the lithology help to explain why correlations of weathering rates with runoff and temperature are often found from long-term monitoring of a single catchment [Gíslason et al., 2009], but that at a global scale these correlations are often not observed.

4.4 Conclusions

Determination of annual element weathering fluxes from the Damma glacier catchment by a variety of methods indicated that using an average annual concentration leads to an overestimation of the annual flux by nearly a factor of two. Particularly in small catchments, where seasonal variations can be quite large, frequent sampling with a frequency of at least once a month is needed to capture the temporal variability in element concentrations. Automated conductivity measurements can provide high resolution information but the relationship between conductivity and individual element concentrations may break down at low and high flow rates. Thus frequent sampling is still required to cover the anticipated range of hydrological conditions.

A compilation of glaciated and non-glaciated granitic and basaltic catchments has revealed fundamental differences in the response of these two silicate lithologies to glaciation events, reinforcing the fact that lithology is a prime control of the response of silicate weathering to climate. Glaciation had no impact on the chemical composition and fluxes (total cations and silica) of stream water draining basaltic catchments. In granitic terrain, glaciation only impacted the fluxes of rivers draining Greenlandic glaciers, resulting in high total cation to silica fluxes, and this is most likely due to a combination of the thermal regime of the glacier and the elevated pH of these rivers. The majority of non-glaciated catchments included in this study were located in mountainous locations which had been previously subject to cold weather processes (glaciation, frost-shattering), obscuring any differences between these and the alpine glaciated catchments. The ratio of total cation to silica fluxes is likely controlled by the amount of erosion in the catchment of which glaciation is one of several enhancing factors. All presently glaciated granitic catchments had clearly elevated K concentrations and slightly elevated Ca concentrations compared to non-glaciated catchments, most likely due to enhanced incongruent weathering. The differing response to erosional regime and climatic parameters between granite and basalt could be due to the texture of the rocks, where the grain-size controls the sensitivity to physical erosion and climate e.g. hydrological regime. Coarse-grained granite is more sensitive to these parameters than fine-grained basalt. Thus, the perturbation of silicate weathering rates and runoff composition due to glaciation will be most pronounced for granitic terrain.

Chapter 5

Calcium isotopes in a proglacial weathering environment: Damma glacier, Switzerland*

5.1 Introduction

The weathering of silicate rocks is thought to influence climate change over geological timescales as the reaction of atmospheric CO_2 with silicates releases divalent cations and bicarbonate into rivers which will ultimately form calcium (Ca) or magnesium carbonates in the ocean, resulting in a net removal of CO_2 from the atmosphere [Berner et al., 1983, Walker et al., 1981]. Yet the amount of silicate weathering now and in the past is poorly quantified, even though it is an important parameter in modern climate models [e.g. Ludwig et al., 1999]. Silicate weathering in the past is typically inferred from proxies such as the strontium isotopic composition of carbonate sediments [Richter et al., 1992]. The riverine Ca flux is a key component of the mass balance of Ca in the oceans and is thought to control short term fluctuations in the Ca isotopic composition of the ocean [Farkaš et al., 2007]. To understand past fluctuations in the Ca isotopic composition of the ocean one needs to ascertain the modern controls of the riverine Ca isotopic flux. The amount of silicate weathering has been assessed by analysing the dissolved load of the largest rivers draining the world's continents [Meybeck, 1987]. This riverine flux will not only consist of silicate sources, but also carbonate and evaporite sources. To distinguish the contribution of each of these sources to the total dissolved ion load, elemental and isotopic ratios are often used [e.g. Gaillardet et al., 1999]. However, this calculation is hampered by the fact that the sources can have similar elemental and

---
*A modified version of this chapter has been published in *Geochimica et Cosmochimica Acta*: R.S. Hindshaw, B.C. Reynolds, J.G. Wiederhold, R. Kretzschmar and B. Bourdon. Calcium isotopes in a proglacial weathering environment: Damma glacier, Switzerland. *Geochim. Cosmochim. Acta*, 75: 106-118, 2011.

isotopic compositions, e.g. carbonate rocks can become enriched in radiogenic strontium (Sr) from coexisting silicate rocks during metamorphism [Palmer and Edmond, 1992]. To characterise the contribution of silicate rocks, it is necessary to analyse well-defined catchments which only drain silicate lithologies. Here, parameters such as vegetation cover and precipitation are known, allowing the actual silicate weathering process for a specific set of conditions to be studied [White and Blum, 1995, Oliva et al., 2003]. This approach has not been without problems since even in granitic catchments, non-silicate minerals such as apatite and calcite may exist [White et al., 1999a].

Despite the dominant role of Ca in modulating the carbon cycle and its importance in biogeochemical cycles, the behaviour of Ca has, in the past, been inferred from measurements of Sr isotopes in conjunction with Sr/Ca ratios [e.g. Capo et al., 1998]. Strontium was chosen because of its similar chemical behaviour to Ca and the ease with which radiogenic Sr variations can be measured. However, Sr may not always be a suitable analogue for the behaviour of Ca, especially where biological cycling is present [Poszwa et al., 2000]. In recent years, however, the techniques have become available which allow the measurement of Ca isotope ratios with suitable precision. Like Sr, Ca can have stable and radiogenic isotopic variations. The radiogenic variations are caused by the decay of ^{40}K to ^{40}Ca, with a decay constant of 5.543×10^{-10} yr^{-1} [Steiger and Jäger, 1977]. Due to the natural dominance of ^{40}Ca, radiogenic enrichments are relatively small and only old rocks with high K/Ca ratios will show significant enrichment [Marshall and DePaolo, 1982]. Thus unlike Sr, where the main applications have used the radiogenic component, Ca isotopic studies have mainly focussed on stable isotope variations [e.g. DePaolo, 2004].

Calcium isotopes have been employed in an increasing number of applications as reviewed by DePaolo [2004], these include: K/Ca dating [e.g. Marshall and DePaolo, 1982], paleo-oceanography [e.g. Farkaš et al., 2007], biological fractionation [Skulan et al., 1997, Bullen et al., 2004, Page et al., 2008, Holmden and Bélanger, 2010], calcite/aragonite precipitation [Gussone et al., 2003, Lemarchand et al., 2004], abiotic fractionation [Ewing et al., 2008] and weathering studies [Schmitt et al., 2003a, Wiegand et al., 2005, Tipper et al., 2006b, 2008a, Cenki-Tok et al., 2009]. These previous weathering studies have been conducted in quite complex settings. The studies in the Strengbach catchment (France) [Cenki-Tok et al., 2009] and Hawaii [Wiegand et al., 2005] highlighted the complicated effects of biological cycling and throughfall on Ca isotope ratios in forested ecosystems and the studies in the Himalaya [Tipper et al., 2006b, 2008a] were conducted in lithologically mixed catchments where secondary precipitation of Ca also occurred. These complex settings made it hard to identify the processes causing the observed Ca isotopic variations.

The objective of this study was to investigate Ca isotopic fractionation during silicate weathering and subsequent biogeochemical cycling by focussing on a lithologically homogeneous alpine catchment with minimal vegetation cover. This simplified the number of potential processes causing isotopic variability. Due to glacial retreat, a soil chronosequence exists in the chosen catchment, allowing both the initial stages of weathering and accompanying soil formation to be studied.

5.2 Sample Collection and Results

5.2.1 Rock, Soil and Mineral Separates

Soil and rock samples were collected as described in section 2.3 and appendix A.1.1 and the procedure used to obtain mineral separates is described in de Souza [2007]. Details concerning the purification and measurement of Ca isotopes can be found in appendices A.3.1 and A.4 and the measurement of element concentrations is described in appendix A.2.1. Major element and calcium isotopic compositions are presented for soils (Table 5.1), rocks and mineral separates (Table 5.2).

Stable $\delta^{44/42}Ca$ rock values ranged from +0.35 to +0.54‰ with an average of +0.44‰ (Table 5.1). This agrees very well with a previous measurement of silicate rock in the Himalayas of +0.45‰ [Tipper et al., 2006b]. The Aar granite has a Rb-Sr dated intrusion age of 298 Ma [Schaltegger, 1994], so it would be expected that K-feldspar could have acquired a radiogenic anomaly equal to 16.5 epsilon. However, the measured epsilon value was 2.6 (Table 5.2), which indicated that resetting had occurred [Dempster, 1986]. Thus, we could not directly measure specific mineral dissolution by tracing the radiogenic Ca component, due to the young K-Ca age of the rocks and the large errors associated with measuring $\epsilon^{40}Ca$. The high Ca content (Table 5.2) of the biotite separate is likely due to inclusions of a Ca-rich accessory phase such as epidote or apatite. These were intergrown with the biotite and were thus unable to be separated during hand-picking.

The major element chemistry of the soils is very similar both between the two sampled horizons and compared to bulk rock values (Fig. 5.1, Tables 5.2 and 5.1). The average isotopic composition of the soil ($\delta^{44/42}Ca$ +0.44‰) remained constant and within error of the range of rock values ($\delta^{44/42}Ca$ = +0.35 to +0.54‰) along the 150 year timespan of the chronosequence (Fig. 5.2). Between the two soil horizons there was no significant isotopic difference (42% probability of 95% significance using a Monte Carlo (MC) t-test with an external reproducibility of 0.07‰, as described in de Souza et al. [2010]). The two much older reference sites were not significantly different in Ca isotopic composition from those within the glacier forefield (6% probability of 95% significance, MC t-test), although these soils were depleted in Ca compared to the forefield soils (Fig. 5.1).

5.2.2 Sequential Extractions

Bulk soil analyses only yield information about average soil chemical and isotopic compositions, whereas sequential extraction allows different soil pools within the bulk

Table 5.1: Age, major element and Ca isotopic composition of the soils analysed for this study. Ca isotope measurements are relative to SRM 915a.

Sample	Age* (years)	[Ca]	[Mg]	[Na] (g/kg)	[K]	[Al]	[Ti]	[Sr] (mg/kg)	$\delta^{44/42}Ca^1$ $2\sigma_{ext} = 0.07$
Soils 0-5 cm									
BL1a	8	9.65	7.42	12.1	30.3	58.9	2.13	170	0.35
BL2a	7	7.85	6.51	13.5	32.9	58.8	1.80	143	0.48[b]
BL4a	12	9.37	4.29	20.7	30.9	63.7	1.78	162	0.48
BL5a	59	12.4	6.82	18.5	26.9	63.5	2.48	199	0.43
BL5a^2									0.45
BL7a	62	10.0	2.68	21.1	30.1	62.9	1.54	178	0.32
BL10a	68	14.6	8.05	15.6	28.4	61.5	3.13	212	0.51
BL10a^2									0.32
BL12a	73	11.3	10.0	12.4	27.4	61.1	2.79	176	0.40
BL13a	76	11.0	10.4	13.7	29.6	62.2	2.69	173	0.28
BL16a	78	10.7	4.13	18.5	26.2	59.4	1.73	188	0.50
BL16a^2									0.35
BL17a	111	12.7	7.16	16.6	26.6	61.4	2.65	194	0.41[a]
BL18a	118	9.50	3.38	17.8	25.4	55.7	1.34	166	0.49
BL20a	129	11.0	4.95	15.1	27.8	57.8	1.82	182	0.43
BL21a	137	10.0	6.81	15.4	26.4	63.2	2.35	194	0.48
BL21a^2									0.33
BL23a	>500	7.53	3.18	11.8	20.7	49.0	2.80	148	0.42
BL24a	>500	6.15	4.98	9.17	19.6	50.8	3.24	145	0.37
BL24a^2									0.37
Soils 5-10 cm									
BL1b	8	9.60	4.60	18.0	30.2	63.0	1.69	176	0.48
BL2b	7	8.68	4.68	18.4	32.1	61.5	1.74	155	0.42
BL2b^3		8.10							0.43[c]
BL4b	12	9.32	4.89	17.0	31.1	60.8	1.68	166	0.52
BL5b	59	10.5	4.38	21.4	28.2	63.6	1.72	197	0.46[a]
BL10b	68	12.1	7.80	17.4	28.9	63.1	2.63	187	0.57[a]
BL12b	73	9.79	9.75	15.4	31.5	65.7	2.33	171	0.44
BL16b	78	10.7	3.91	19.5	28.4	61.5	1.77	190	0.53[a]
BL17b	111	12.0	5.77	18.6	26.6	62.4	2.34	191	0.52
BL18b	118	10.3	2.69	19.4	27.9	60.9	1.37	182	0.38
BL21b	137	9.93	9.76	16.2	29.5	67.8	2.62	186	0.51[a]
BL23b	>500	7.47	3.81	13.4	22.6	55.4	2.77	156	0.45
BL24b	>500	6.90	5.43	10.7	23.7	60.3	3.29	171	0.41[a]
BL24b^3		6.66							0.46
Additional Soils^4									
S01 glacial sand	0	5.19	2.14	12.6	27.7	47.3	0.87	88	0.35
S05 glacial sand	10	10.2	8.52	13.3	26.7	54.8	2.35	155	0.39[b]
S06 root zone	10	9.26	9.34	12.7	26.0	53.7	2.21	141	0.40
S04	14	10.4	10.5	12.3	28.7	60.9	2.59	191	0.43
S07	55	10.2	6.65	16.5	26.1	59.7	2.07	180	0.47
S08 root zone	55	10.0	5.74	18.0	25.9	62.9	1.86	186	0.43
S08 mineral soil	55	9.83	6.48	16.7	27.0	62.4	1.95	179	0.43
S13 Oxyria root zone	56	8.53	4.99	15.0	33.3	60.4	1.73	143	0.39
S13 Rhododendron root zone	56	8.47	5.24	15.1	33.6	60.9	1.48	143	0.34[a]
S09	66	9.46	7.23	14.6	25.9	58.6	1.90	179	0.50
S10 Organic-rich	70	10.2	15.2	8.74	34.5	65.9	3.72	147	0.55
S10 Mineral soil	70	11.3	14.2	10.5	34.7	68.9	3.63	170	0.48
S11 Organic-rich	75	10.3	8.64	7.09	22.5	47.9	2.60	129	0.48
S11 Mineral soil	75	12.5	12.8	10.8	32.0	65.3	3.49	187	0.54[a]
S12 Organic-rich	80	10.2	10.0	8.22	24.3	47.7	2.76	142	0.45

*Age in 2008
^{1} Average of n=2 unless otherwise indicated, [a] n=1 [b] n=3 [c] n=10
^{2} repeat analysis approximately one year after first measurement
^{3} complete procedural replicate
^{4} Elemental concentration data and sample names from de Souza et al. [2010]

soil to be investigated. The sequential extraction procedure is described in appendix A.2.2. Three soils were chosen (0-5 cm depth); a very young soil (Soil BL4a, ~10 years old), a soil mid-way down the chronosequence (Soil BL17a, ~110 years old) and the

Table 5.2: Analyses of rock, soil and plant samples.

Sample[1]	[Ca]	[Mg]	[Na]	[K]	[Al]	[Ti]	[Sr]	$\delta^{44/42}Ca$[2]	$\epsilon^{40}Ca$[2]
			(g/kg)				(mg/kg)	$2\sigma_{ext} = 0.07$	
Rocks									
R02	14.9	4.03	14.0	30.1	56.8	2.65	247	0.47	-0.8
R03	7.74	9.82	13.9	33.8	66.0	2.67	156	0.35	-1.2
R04	13.1	6.35	15.1	31.8	69.0	2.29	285	0.44	-1.1
R05	3.02	-	16.3	34.6	54.5	0.22	40	0.54	1.9
R06	6.76	10.2	15.3	30.0	65.8	2.53	177	0.36	-1.5
R07	6.88	1.90	14.9	40.4	56.0	0.90	74	0.45	0.1
R08	14.4	7.54	13.8	33.2	67.1	3.30	270	0.47	0.5
Average rock								0.44	-0.3
Minerals									
R07 Biotite	20.7	16.7	5.02	40.6	43.3	58.1	54	0.50	0.0
R07 Plagioclase	4.95	0.33	31.7	8.03	48.5	0.64	62	0.47	-0.1
R07 K Feldspar	1.00	0.06	20.3	82.8	96.8	0.01	38	0.36	2.6
Plants[3]									
RhL_Y_6_08	0.99	1.15	0.02	13.1	0.00	0.00	1	-0.23	2.9
RhL_1M_7_08	4.07	1.44	0.00	7.39	0.02	0.00	5	-0.19	1.3
RhL_2M_8_08	5.93	1.57	0.00	6.04	0.05	0.00	9	-0.22	0.7

[1] Sample labels as given in de Souza et al. [2010]
[2] Average of 2 measurements except for average soils.
[3] Element contents based on dry weight

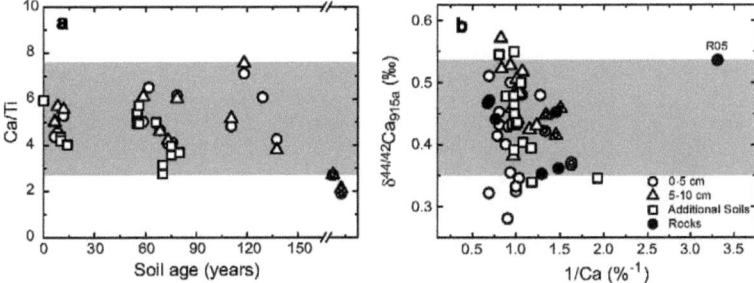

Figure 5.1: (a): The Ca/Ti ratio variation over time. This ratio remains constant, only decreasing in the reference soils. The shaded area in both figures represents the corresponding range of rock values (In (a), the Ca/Ti ratio of R05 (13.5) is omitted). (b): Plot of $\delta^{44/42}Ca$ against 1/Ca illustrating that although heterogeneous, the isotopic compositions of the soils can be explained by mixing between different rock samples. Rock R05 appears to be an outlier. The reference soils are points marked with internal circles.

reference site (Soil BL24a, >500 years old).

Table 5.3 shows that total elemental recoveries, calculated by adding up the amount of each element extracted in each step compared to the bulk soil digest, were for the most part within ± 20%. One exception was the recovery of K in Soil 4, this was probably due to an incomplete digest of the bulk soil. A ternary plot with Ca, Mg and Na+K at the vertices reveals how closely the extraction procedure reached the target pools, and highlights the variability between soils of different ages (Fig. 5.3).

Calcium was the most dominant cation in the small exchangeable pool. This pool increased in size for Ca from 1-2 to 5% in 100 years. The organically bound fraction (step 2) contained nearly 20% of the Ca along with phosphate and so it is likely that this step is

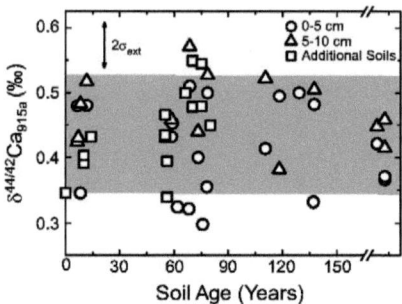

Figure 5.2: Ca isotopic composition of the soils as a function of time. The shaded area illustrates the range in $\delta^{44/42}$Ca of the rock.

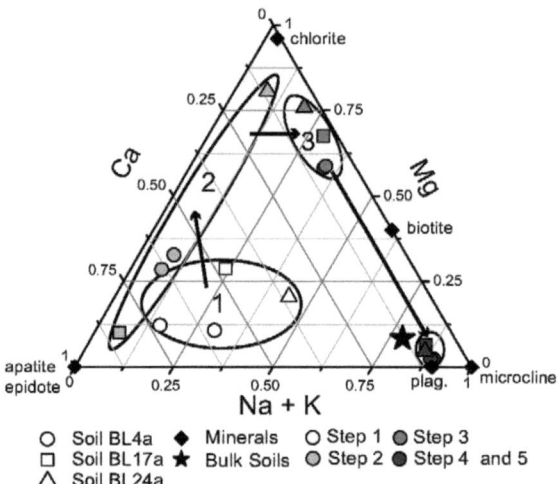

Figure 5.3: Ternary plot of the sequential extraction data using normalised molar concentrations. To allow comparison with Soil BL17a the last two extraction steps for the other soils are combined. Soil BL4a was analysed twice and the three bulk soils plot in the same place so only a single point is observed. Soil BL24a (reference site) is clearly different from the other two soils with respect to the first two steps.

dissolving apatite or organically bound P and Ca. Magnesium and Fe were also released, possibly indicating dissolution of secondary (hydr)oxides. Step 3 attacked phyllosilicates (biotite, chlorite) and this extraction step exhibited the highest concentrations of Mg and Fe, for which the main host minerals in this catchment were biotite and chlorite. Where there were five steps in the procedure, it is clear that the fourth step has not been able to fully break down the most resistant silicates - feldspars and plagioclase. It was only with the use of HF in the microwave digest that the residual phase was totally dissolved, releasing large amounts of Na and K as the most resistant minerals dissolved. The aqua regia dissolution (Step 4) was, however, where the highest concentrations of Ca were found.

The isotopic values of the individual sequential extraction fractions can be summed using the formula below and compared to the bulk soil digest isotopic value to check for isotopic mass balance.

$$\delta^{44/42}Ca_{sum} = \sum_{i=1}^{n} f_{Ca,i}\delta^{44/42}Ca_i \qquad (5.1)$$

where i was the sequential extraction step of n steps, $f_{Ca,i}$ was the fraction of Ca extracted in step i and $\delta^{44/42}Ca_i$ was the $\delta^{44/42}Ca$ value of the extracted fraction. Assuming that the error on $\delta^{44/42}Ca$ dominates (0.07‰) this leads to a combined error of 0.16 ‰ for the 5-step procedure. Thus, the bulk soil digest Ca isotopic values are within error of those calculated using the above formula. For the two soils in the main chronosequence (BL4a and BL17a), $\delta^{44/42}Ca$ values were lower than bulk soil in the first extraction step (exchangeable pool). In Soil BL4a there was a pronounced increase in $\delta^{44/42}Ca$ from the first to the third extraction step (Fig. 5.4). There was no significant variation between the different pools extracted in Soil BL24a (reference site).

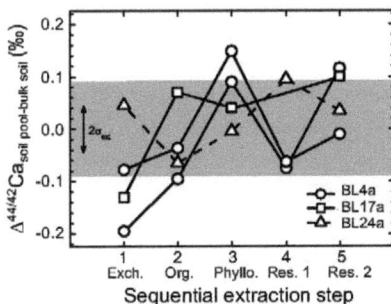

Figure 5.4: Ca isotope composition of the different sequential extraction steps (see Table A.3) for three different soils. In order to compare the sites, all were normalised to the bulk soil isotopic composition of that site. The sequential extraction procedure was carried out in duplicate for Soil BL4a. Values are reported as $\Delta Ca_{soilpool-bulksoil}$ (where the soil pool is a particular sequential extraction step e.g. exchangeable) and the combined external error is indicated. The shaded area highlights the region within which the points are within error of the bulk soil.

Table 5.3: Elemental and Ca isotopic compositions of soil sequential extractions

Sample	Age (years)	Extraction Step	[Ca]	[P]	[Mg] (mmol/kg)	[Fe]	[Na]	[K]	Ca fraction of soil (%)	$\delta^{44/42}$Ca [1] $2\sigma_{ext} = 0.07$
BL4a-1	12	1	2.49	0.17	0.42	0.09	0.09	0.45	1	0.29
		2	30.4	**13.6**	13.6	14.3	0.60	3.18	17	0.39
		3	16.3	5.44	**120**	**137**	1.77	67.9	9	0.57
		4	**117**	0.12	34.2	67.9	1.27	39.0	66	0.41
		5	11.9	0.07	2.38	40.7	**817**	**532**	7	0.60
		sum	*178*	*19*	*170*	*260*	*821*	*642*		*0.43*
BL4a-2	12	1	2.73	0.02	0.50	0.11	0.45	0.95	2	0.40
		2	30.5	**14.4**	17.1	19.2	0.67	3.97	17	0.44
		3	14.1	2.16	**114**	**128**	2.03	64.3	8	0.63
		4	**123**	0.04	34.2	67.6	1.49	43.5	68	0.42
		5	10.6	0.05	1.85	36.7	**740**	**521**	6	0.47 [b]
		sum	*181*	*17*	*168*	*252*	*745*	*634*		*0.45*
BL4a	12	bulk	197	12	143	210	886	353		0.48
BL17a-1	111	1	3.12	0.15	1.90	0.34	0.36	1.22	5	0.28
		2	34.3	**20.7**	4.12	11.6	0.98	1.64	15	0.48
		3	2.98	1.38	57.8	52.4	1.22	23.6	9	0.45
		4	**160**	0.40	**129**	**184**	**1064**	**614**	71	0.51
		sum	*201*	*23*	*193*	*248*	*1066*	*640*		*0.49*
BL17a	111	bulk	165	23	227	281	1234	541		0.41 [a]
BL24a-1	>500	1	4.67	0.42	2.70	0.45	1.01	4.74	5	0.42 [a]
		2	2.58	24.6	18.8	44.0	0.13	1.79	3	0.31
		3	5.26	**25.6**	**96.8**	**143**	1.09	24.4	5	0.37
		4	**100**	1.50	42.9	80.5	2.69	80.7	81	0.47
		5	10.3	0.32	12.5	47.3	**556**	**379**	7	0.40
		sum	*123*	*52*	*174*	*315*	*561*	*490*		*0.45*
BL24a	>500	bulk	138	43	158	288	893	408		0.37

Bold type indicates in which fraction the element was most concentrated
[1] Average of n=2 unless otherwise indicated [a] n=1 [b] n=4

5.2.3 Water Samples

The Ca isotope values of the stream water at all three sites (described in section 2.4) remained essentially constant, with an average value of $\delta^{44/42}$Ca= +0.48‰ and with no obvious seasonal variation, whilst the Ca flux displayed large seasonal changes (Fig. 5.5). The Ca isotopic composition of the water samples was not significantly different from that of the soils (29% probability of 95% significance, MC t-test) (Fig. 5.6). However, taken on its own, the 24th June sample exhibited a significantly heavier (76% probability of 95% significance, MC t-test) isotopic composition compared to the soils (Fig. 5.5). Snow samples had lower Ca concentrations and larger $\delta^{44/42}$Ca values compared with rainwater samples (Table 5.4). Snow and rainwater isotopic values were similar to Cenki-Tok et al. [2009]. This is to be expected since the study areas are geographically close.

5.2.4 Plants

Leaves from *Rhododendron ferrugineum* were collected during the growing season from an ~75 year old site. Each sample is a composite sample of leaves from different specimens. Leaf samples were enriched in light Ca isotopes compared to bulk soil (Fig. 5.6), the exchangeable pool (Table 5.3) and the porewaters (Table 5.4). This yielded a

Sample Collection and Results

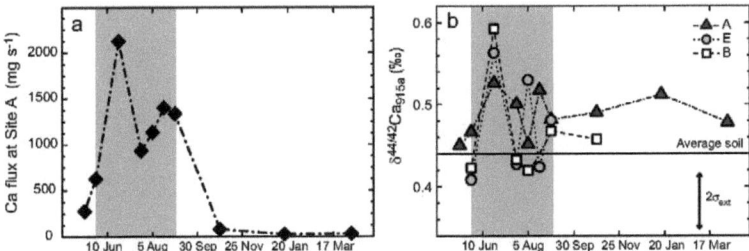

Figure 5.5: (a) Calcium flux at Site A. The shaded area in both graphs marks the main melt season. (b) Isotopic composition of the three stream water sampling sites (see Fig. 2.3) as a function of time. Average soil Ca isotopic composition is marked for comparison.

Table 5.4: Calcium isotopic composition and concentration data of the water samples

Site	Date (YMD)	Discharge[1] (L s^{-1})	[Ca] (μg L^{-1})	Flux mg s^{-1}	$\delta^{44/42}$Ca[2] $2\sigma_{ext} = 0.07$
Stream water					
A	20080513	*250*	1103	276	0.45
A	20080527	*800*	780	624	0.47
A	20080624	4502	472	2126	0.53
A	20080722	1908	487	930	0.50
A	20080805	3595	315	1133	0.45
A	20080819	4110	340	1398	0.52[b]
A	20080902	4221	316	1335	0.48
A	20081028	84	974	82	0.49
A	20090116	30	1357	27	0.51
A	20090408	30	1694	34	0.48
B	20080527		641		0.42
B	20080624		439		0.59[a]
B	20080722	421	442	186	0.43
B	20080805	2092	262	549	0.42
B	20080902	2078	267	555	0.47
B	20081028		762		0.46
E	20080527		800		0.41
E	20080624		442		0.56
E	20080722	892	414	369	0.43
E	20080805	1866	295	551	0.53
E	20080819	1083	413	447	0.42
E	20080902	723	440	318	0.48
Groundwater					
G1	20080930		931		0.58
G1	20080930[3]		931		0.48
G2	20080930		1999		0.45
Porewater					
hm92-1			6092		0.45[a]
EP2			3018		0.46
BR2			18132		0.31
Precipitation					
rain	20080708		508		0.46
rain	20080916		662		0.36
rain	20080930		768		0.36
snow	20080513		87		0.49
snow	20081028		248		0.58

[1] discharge data in italics is estimated where possible (see text)
[2] Average of n=2 unless otherwise indicated, [a] n=1 [b] n=3
[3] repeat

$\Delta^{44/42}$Ca$_{leaf-bulksoil}$ of ≈ -0.64‰. The preferential uptake of light Ca by plants is in agreement with previous studies [e.g. Wiegand et al., 2005]. The Ca isotopic composition of plants in this forefield is described in more detail in chapter 6.

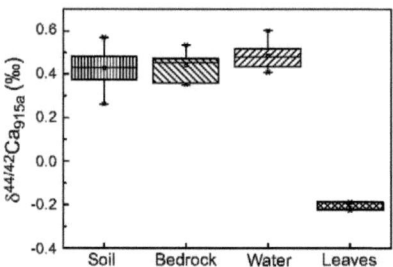

Figure 5.6: Box plot summarising the main pools of calcium analysed. Leaves are from *Rhododendron ferrugineum*. This is the only significantly fractionated bulk pool of Ca in the forefield.

5.3 Discussion

5.3.1 Weathering in soils

This field site contains young soils (Hyperskeletic Leptosols), which are poorly developed. The chemical index of alteration ($Al_2O_3/(Al_2O_3+CaO+Na_2O+K_2O)$) of the soils (0.62) was identical to that of the rocks (0.61) and ratios of mobile to immobile elements indicated lack of significant leaching from the soils (Fig. 5.1a). Secondary clay formation resulting from weathering was only detectable by X-ray diffraction (XRD) analysis at the sites older than ~100 years. Earlier formation of small quantities of clay, poorly crystalline secondary phases which are not detectable by XRD, or the alteration of rocks without formation of secondary phases [Miller and Drever, 1977] could not be ruled out. The isotopic composition of the soils are determined by mixing between slightly different initial rock compositions without significant mass dependent fractionation (Fig. 5.1). The coincidence of soil and rock $\delta^{44/42}Ca$ values was also reported in the Himalaya [Tipper et al., 2006b].

In order to gain more information about the potential sources of Ca to the dissolved load we performed sequential extractions on bulk soils. The largest soil pool of Ca is in the residual fraction which is dominated by plagioclase and feldspar (Fig. 5.3). There is no decrease in the Ca concentration of this resistant fraction over time, indicating that these Ca rich phases are not yet significantly affected by weathering (even in the reference soil, BL24). When the dissolution of the residual phase was split into two steps, the highly resistant silicates, sodic plagioclase and microcline, did not appear to break down until Step 5, as indicated by the high concentrations of Na and K (Table 5.3). Although the plagioclase is close to the albite endmember, it is known that plagioclases exhibit zoning with Ca and Na rich parts and that the Ca rich endmember is more easily weathered

[Oliva et al., 2004]. Thus, the high concentrations of Ca in Step 4 could stem from a calcic plagioclase (anorthite) or epidote. An extended X-ray absorption fine structure (EXAFS) spectroscopy study on Fe in the forefield found that the Fe fraction within epidote remained constant over the chronosequence, indicating lack of epidote weathering [Kiczka-Cyriac, 2010]. This finding, together with the fact that at pH 5 the weathering rate of anorthite is several orders of magnitude faster than that of epidote [Lasaga et al., 1994], suggests that anorthite is being released in Step 4 with Step 5 dissolving albite and epidote. The apparent stability of epidote in these soils is in contrast to a recent study, based on XRD data, which reported a rapid decrease of epidote content over a 150 year time period in a similar alpine chronosequence [Mavris et al., 2010].

The 'organically bound' fraction (Step 2, Table 5.3) comprised the second largest pool of Ca in the two forefield soils. These sites contain little organic matter [Smittenberg et al., 2009], thus we do not think that the Ca is predominantly bound to organic matter. Rather, the dominance of P and Ca in this fraction, strongly points to the influence of apatite dissolution. Soil BL17a released Ca and P identical in proportion to the ratio in apatite (5:3), whilst BL4a released slightly more Ca than expected based on congruent apatite weathering. Laboratory experiments have shown that during initial apatite dissolution, weathering is incongruent, resulting in higher initial Ca/P ratios [Valsami-Jones et al., 1998]. The absence of Ca in the second extraction step at BL24a could indicate that all the apatite has been weathered and the Ca has been removed, either by leaching or by plant uptake [Blum et al., 2002]. Apatite depletion has previously been reported in young soils (Nezat et al. [2008] and references therein) and could provide a vital source of Ca for plants at the youngest sites. A pure sample of apatite from our fieldsite to compare with the isotopic composition of this extraction step could not be obtained, but based on the isotopic composition of this extraction step we would expect the isotopic composition of apatite to be similar to whole rock values. Thus, we do not expect the weathering of apatite to shift the $\delta^{44/42}Ca$ value of the exchangeable pool away from that of the residual soil pool.

5.3.2 Porewaters and the exchangeable pool

Rapid equilibration between porewaters and the exchangeable pool is often assumed to occur [Drever, 1997] and would result in identical $\delta^{44/42}Ca$ values in these two pools. Although only three porewater values were measured, it is clear that spatial heterogeneity exists (Table 5.4). A comparison can be made between the porewater BR2 and Soil BL17 which were located within 50 m of each other; the $\delta^{44/42}Ca$ values were +0.28‰ (exchangeable) and +0.31‰ (porewater). Both values are lower than bulk soil ($\delta^{44/42}Ca$ at Site BL17 is +0.41‰). It is, nevertheless, hard to tell whether porewater $\delta^{44/42}Ca$ controls the exchangeable pool or vice versa, or indeed whether it is realistic to assume that spatial homogeneity exists on this scale. Whichever scenario is assumed, a pool of light Ca is present at the inception of soil formation.

The small size of Ca in the exchangeable pool (1-5% of the total soil Ca) could make it susceptible to external inputs or vegetation. The $\delta^{44/42}$Ca of precipitation (dry and wet combined) measured at this site was identical to the $\delta^{44/42}$Ca of rocks and soils, thus if there is an effect of precipitation on the exchangeable pool it will be hard to observe. Vegetation could potentially control porewater and exchangeable $\delta^{44/42}$Ca values [Holmden and Bélanger, 2010, Bullen et al., 2004]. At Damma, fractionation of the exchangeable pool or porewaters is unlikely to be caused by vegetation, since the size of the Ca biomass pool in the glacier forefield is, on average, only $\sim$2% of the size of the exchangeable Ca pool in the 0-5 cm soil layer.

Low $\delta^{44/42}$Ca values of the exchangeable fraction (Step 1, Fig. 5.4) could be explained by the preferential adsorption of light Ca by soil components. The $\triangle^{44/42}$Ca$_{soilpool-bulksoil}$ of the exchangeable fraction increased from BL4a to BL24a (Fig. 5.4). We use cation exchange capacity (CEC) measured using 1 M NH$_4$Cl as a proxy for the number of sorption sites available. From BL4a to BL24a the size of the exchangeable Ca pool increases by 250% (Table 5.3) and the CEC increases by 850% (6.4 - 59.1 mmol$_c$kg^{-1}), representing an increase in organic matter and clay minerals. Thus, with increasing soil age, there will be a greater variety of sorption sites available. Different sorption sites (e.g. organic vs. mineral) could potentially have different fractionation factors [Lemarchand et al., 2005] and this could change the $\triangle^{44/42}$Ca$_{exchangeable-bulksoil}$ depending on which sites Ca sorbs to.

5.3.3 Stream water

The similarity of the stream water Ca concentration and $\delta^{44/42}$Ca to rain water (Table 5.4) could potentially be due to the stream being dominated by rain water. The Ca isotopic composition of the precipitation was not constant throughout the year. In particular, snow has a higher $\delta^{44/42}$Ca value (average = +0.53‰) than rain (average = +0.39‰) and is closer to the seawater composition ($\delta^{44/42}$Ca = +0.96‰). The percentage contribution of precipitation (rain, snow and ice melt) to Ca in the river was calculated using a hydrograph separation model for a neighbouring catchment [Verbunt et al., 2003] combined with a conventional chloride correction using concentrations measured in the Damma catchment. This contribution varied from 2% in the winter months to nearly 35% at the height of summer. Although this correction reduced absolute Ca concentrations, $\delta^{44/42}$Ca values remained, within error, unchanged. Thus, the isotopic composition of rain is close enough to that of the silicate rock in this catchment to not affect the resultant Ca isotopic composition of the river.

The average $\delta^{44/42}$Ca of the stream was, within error, identical to soils and rocks (Fig. 5.6) indicating that no significant fractionation occurred during the dissolution of primary silicate minerals. The calcium isotopic composition of the stream showed no significant correlation with discharge at Site A, in agreement with the results of Cenki-Tok et al. [2009] and Holmden and Bélanger [2010], even though Ca concentrations varied by a factor of five over the season.

The heaviest dissolved Ca isotopic compositions, coincident at all three sampling sites, were recorded on 24th June - when the discharge reached its maximum after the snowmelt period. Increased snowmelt contribution was invoked to explain the highest $\delta^{44/42}$Ca value recorded in the Strengbach catchment [Cenki-Tok et al., 2009]. The high $\delta^{44/42}$Ca value of these water samples could be caused by (1) the isotopic composition of the snow itself, (2) snow melt flushing out biological degradation products or (3) snow melt flushing out a pool influenced by secondary processes.

If we assume that 'old' snow contributes most to the isotopic composition of snowmelt, then spring snow melt with the May Ca isotopic composition, which is similar to rock, cannot be the cause of the high $\delta^{44/42}$Ca value observed on 24th June.

Alternatively, snow melt could be flushing out a pool of heavy Ca, which had not been in contact with baseflow during the winter. This pool could be due to biological degradation products which have accumulated over the winter [Clow and Mast, 2010]. However, as previously stated, biological control is unlikely due to the small size of the biological pool. Degrading organic matter would be expected to return light Ca back into the system (since plants are isotopically lighter than bulk soils, Table 5.2), resulting in a decrease in $\delta^{44/42}$Ca during snowmelt which is opposite to that observed.

We favour the third explanation: that secondary processes, namely adsorption (including ion exchange), induced by winter hydrology could cause increased $\delta^{44/42}$Ca values which are then flushed out during snowmelt. In winter, when the vast majority of channels freeze, isolated pockets of water could develop high $\delta^{44/42}$Ca values due to preferential sorption of light Ca onto soil components and when discharge and channel connectivity increase during snowmelt the residual heavy Ca is flushed out. In summer, the braided channel network ensures continual flushing occurs and prevents isolated pockets of water forming.

Sorption of light Ca onto soil components is, as far as we are aware, not yet documented but fractionation during adsorption has been observed for several other stable isotope systems [Balistrieri et al., 2008, Juillot et al., 2008, Mikutta et al., 2009]. The seasonal precipitation and dissolution of secondary mineral phases (to which Ca could adsorb) has previously been reported in an alpine catchment [Clow and Drever, 1996]. Detailed laboratory and field studies are needed to investigate Ca isotope fractionation during adsorption. Similar hydrological control of the isotopic composition of river waters might be applicable to other climates where there are large seasonal variations in water discharge. There are very few seasonal stream water Ca isotopic studies but comparison of high and low stages of monsoonal rivers reveals that there is likely also a seasonal trend in these large rivers [Tipper et al., 2010].

5.4 Conclusions

In this very young weathering environment (0-150 years), no fractionation of Ca isotopes was observed in bulk soil or stream waters compared to the bedrock. Pore and ground waters sampled in the forefield were also unfractionated with respect to bulk rock values. No seasonal variation was observed in the stream, but the hydrological event of snowmelt had a transient impact on streamwater $\delta^{44/42}Ca$, which could be due to seasonal adsorption of Ca. Investigation of the soil pools by sequential extraction indicated that the exchangeable pool is fractionated compared to the parent soil, but the mechanisms causing this are not yet fully understood. Vegetation was significantly enriched in light Ca but vegetal biomass is not yet significant enough to have induced a measurable fractionation on the soils and waters of the catchment.

Our observations imply that Ca isotopes can only become fractionated during silicate weathering if secondary processes are involved (e.g. uptake by vegetation). In future, it will be essential to determine whether such secondary processes play an important role in determining the riverine Ca isotope flux to the world's oceans.

Chapter 6

Calcium isotope fractionation in alpine plants*

6.1 Introduction

The presence of vegetation has a major impact on the biogeochemical cycles of many elements, through uptake, re-cycling and the acceleration of weathering rates [Drever, 1994, Bormann et al., 1998]. As one of the essential plant macronutrients, calcium has been particularly intensively studied [Likens et al., 1998, McLaughlin and Wimmer, 1999, White and Broadley, 2003].

Calcium plays a vital role in plant growth and development. Calcium is used for signalling, maintaining membrane integrity, stomatal regulation, enzyme activation and for the structural integrity of the plant, where it is a major component of cell walls [Marschner, 1995, McLaughlin and Wimmer, 1999]. Uptake of Ca occurs by mass flow of Ca, driven by transpiration, into the vicinity of cells walls, where it is adsorbed, typically by the carboxyl groups of cell wall pectins [Haynes, 1980]. Within the roots, a small fraction of Ca flow is symplastic (through cells) as it is required within cells for signalling, but the majority of flow is non-metabolic [Drew and Biddulph, 1971] and apoplastic (between cells) [White and Broadley, 2003]. The vast majority of Ca transport in the plant is through the xylem and the accumulation of Ca is determined by the transpiration rate and the root cation exchange capacity, both of which are species dependent [White and Broadley, 2003]. Plant leaves are a sink for Ca and unlike most other plant nutrients, Ca is not retranslocated later in the growing season. Calcium is the least mobile of the plant macronutrients in phloem tissue and plants lose Ca through leaf fall and leaching [Marschner, 1995, McLaughlin and Wimmer, 1999].

Calcium uptake by vegetation has previously been investigated using a combination of

---

*A modified version of this chapter has been submitted to *Biogeochemistry*: R.S. Hindshaw, B.C. Reynolds, J.G. Wiederhold, M. Kiczka, R. Kretzschmar and B. Bourdon. Calcium isotope fractionation in alpine plants.

element ratios (typically Ca/Sr and Ca/Ba) and radiogenic Sr [Bailey et al., 1996, Poszwa et al., 2000, Bullen and Bailey, 2005, Dasch et al., 2006, Blum et al., 2008, Drouet and Herbauts, 2008, Pett-Ridge et al., 2009]. The results from these studies show that, in addition to mineral weathering, atmospheric deposition is an important source of Ca for plants [Bullen and Bailey, 2005, Blum et al., 2008, Pett-Ridge et al., 2009]. In densely vegetated environments or in highly weathered soils, the majority of Ca released when a plant dies is taken up by other plants (biologically recycled) and only a small fraction is exported to streams [Bullen and Bailey, 2005, Blum et al., 2008]. Although radiogenic Sr can in theory act as an analogue for Ca [Capo et al., 1998], the sources of Sr and Ca are not exactly the same and the use of elemental ratios is complicated by varying discrimination factors within the plant [Poszwa et al., 2000, Drouet and Herbauts, 2008].

Radioactive ^{45}Ca has a long history of use in plant nutrition studies [Drew and Biddulph, 1971, Ferguson and Bollard, 1976]. However, it is only recently that stable Ca isotopes have been utilised in vegetation studies [e.g. Holmden and Bélanger, 2010]. Stable isotopes have the potential to refine our understanding of the biological cycling of elements and act as tracers for the impact of vegetation in biogeochemical cycles. Within the biogeochemical Ca cycle, uptake by vegetation is one of the few processes known to induce large stable Ca isotope fractionation in nature [Wiegand et al., 2005, Page et al., 2008, Cenki-Tok et al., 2009, Holmden and Bélanger, 2010, Hindshaw et al., 2011a]. Calcium isotopes thus have the potential to trace Ca cycling within ecosystems, but first, the exact processes inducing Ca isotope fractionation between plant tissues and during uptake need to be investigated.

The majority of previous studies have focussed on forested ecosystems, and have clearly shown that within-plant Ca isotope fractionation occurs, with root samples exhibiting the lightest Ca isotopic compositions in the plant (lighter than source Ca) and leaves the heaviest [Wiegand et al., 2005, Page et al., 2008, Holmden and Bélanger, 2010, Cenki-Tok et al., 2009]. It was proposed that ion-exchange reactions with the cell walls surrounding the xylem could cause the observed fractionation effects if the lighter Ca isotopes were preferentially retained [Wiegand et al., 2005, Page et al., 2008]. The 'one-way' flow of Ca, due to its lack of retranslocation, should simplify the understanding of fractionation processes. It is not clear at present what causes the large Ca isotope fractionation between soil and roots. Identifying the source of Ca is complicated in forested ecosystems due to the presence of chemically (and isotopically) distinct soil horizons and different rooting depths [Holmden and Bélanger, 2010]. As different plant species will have different demands for calcium [Marschner, 1995] this will likely lead to species specific fractionation patterns. In addition to species differences, it can be expected that the seasonal growth cycle will also affect Ca isotope fractionation. Recent observations have shown that old needles of a spruce tree had a heavier Ca isotopic composition than younger needles [Cenki-Tok et al., 2009].

Previous studies of Ca isotope fractionation in vegetation have concentrated on tree species. In order to investigate Ca isotope fractionation in a wider range of plant families, we analysed several species of small alpine plants. The small size of alpine plants allowed us to collect complete plant samples and obtain whole plant Ca isotopic compositions, in addition to those of individual tissues. In order to directly relate these results to previous work on Ca isotope fractionation in an alpine environment [Hindshaw et al., 2011a], plant samples were taken from the same fieldsite (Damma glacier). This allowed us to also investigate the effect of soil heterogeneity and mycorrhiza on plant Ca isotopic compositions. The aims of the study were to investigate species differences, within plant fractionation and seasonal differences in plant Ca isotopic compositions.

6.2 Sample description

All plant samples were collected from the Damma glacier forefield (Chapter 2). The preparation of plant samples for analysis is described in appendix A.1.3. Plant samples were analysed for major element concentrations (appendix A.2.1) and Ca isotopes (appendices A.3.1 and A.4). A range of different species from different plant orders were analysed in this study (Table 6.1).

Table 6.1: List of plant species analysed in this study

Full species name	Abbreviated name used in text	Family	Type	Samples analysed		
				PT	L	CR
Rhododendron ferrugineum	Rhododendron	Ericaceae	woody	×	×	
Salix helvetica	Salix h.	Salicaceae	woody			×
Salix retusa	Salix r.	Salicaceae	woody			×
Oxyria digyna	Oxyria	Polygonaceae	herb		×	
Rumex scutatus	Rumex	Polygonaceae	herb	×		
Leucanthemopsis alpina	L.alpina	Asteraceae	herb			×
Agrostis sp.	Agrostis	Poaceae	grass		×	

PT - Different plant tissues
L - Seasonal leaf samples
CR - Above-ground plant samples from along the chronosequence

Three sets of samples were collected from the forefield in 2008. The first set of samples consisted of tissues (root, stem, leaf and flower samples) from individual specimens of *Rhododendron* and *Rumex* to investigate within-plant Ca isotope fractionation. Two different specimens of *Rhododendron* were analysed (RhA and RhB), in addition to the roots of a third (RhC). The complete root systems of the *Rhododendron* specimens were unable to be collected due their great areal extent. RhA and RhB were from soil which was 75 years old, and RhC was from soil which was 110 years old (at the time of sampling in 2008). Three types of soil material surrounding RhB were analysed in order to investigate whether the immediate soil environment was influenced by the plant. The three types of soil samples were: organic soil (uppermost layer below the litter horizon), bulk soil (homogenised soil sample from 0-5 cm below the organic soil layer) and rhizosphere soil (soil which had to be shaken off plant roots). The *Rumex* specimen grew on soil which was approximately 10 years old and here the complete root system was obtained. The root samples of *Rumex* and RhC were peeled to give a stele sample (innermost part of the root) and a cortex sample (outer part of the root). This sample set has previously been described and analysed for iron isotopes by Kiczka et al. [2010].

The second set of samples was used to investigate seasonal variations in plant Ca isotopic composition. These samples were collected from a site where the soil was approximately 75 years old and consisted of leaves from *Rhododendron* and *Oxyria*, and leaves and flowers from *Agrostis*. Leaf samples were collected at one month intervals over the growing season. Each sample was a composite sample of leaves from several individual plants. Due to the difficulty of separating *Agrostis* leaves from the stem, *Agrostis* 'leaf' samples contain both stem and leaf tissue.

The third set of samples served to investigate if the age of the soil affected the above-ground biomass Ca isotopic composition. The above-ground plant parts of *L. alpina*, *Salix h.* and *Salix r.* were collected from soils of different ages along the chronosequence. The infection of plant roots by mycorrhizal fungi may alter the plant Ca isotopic composition. In order to investigate this, the percentage of roots infected by mycorrhizal fungi was estimated microscopically using the gridline intersection method [Giovannetti and Mosse, 1980]. In addition, sporocap (fruiting body) samples from two different species of mycorrhiza were sampled.

Plants acquire Ca from the soil and the Ca isotopic composition of this source pool has been described by Hindshaw et al. [2011a]. The labile (plant-available) pool of Ca in the soil can be represented by the soil exchangeable pool. The soil exchangeable pool was extracted from seven soils along the chronosequence as described in appendix A.2.2. The Ca isotopic composition was measured in four of the extracts (two of these sites (BL4a and BL17a) were previously reported in Hindshaw et al. [2011a]).

Whole plant Ca isotopic compositions were calculated using the following mass balance equation:

$$\delta^{44/42}Ca_{plant} = \frac{\sum_i m_i[Ca]_i \delta^{44/42}Ca_i}{\sum m_i[Ca]_i} \quad (6.1)$$

where i is a plant part (root, stem, leaf, flower), m is the dry mass and $[Ca]$ is the Ca concentration. The root mass of *Rhododendron* was estimated to be 25% of the whole plant mass based on a study of *Rhododendron arboreum* [Rana et al., 1989]. The Ca isotopic composition of the root cortex was not analysed but was calculated by mass balance using the following equation:

$$\delta^{44/42}Ca_{cortex} = \frac{\delta^{44/42}Ca_r[Ca]_r - \delta^{44/42}Ca_s[Ca]_s f_s}{[Ca]_r - [Ca]_s f_s} \quad (6.2)$$

Table 6.2: Sr/Ca, Ba/Ca and $\delta^{44/42}$Ca of the soil exchangeable pool

Soil	Soil age (years)	Sr/Ca mmol/mol	Ba/Ca mmol/mol	$\delta^{44/42}$Ca $2\sigma_{ext}=0.07$‰
BL2a	7	2.9	12.5	
BL4a	12	1.8	9.2	0.34
BL7a	62	1.6	11.8	0.45
BL10a	68	1.7	10.4	
BL13a	76	1.8	6.9	0.52
BL17a	111	2.5	10.6	0.28
BL20a	129	2.0	3.1	
Average		2.0	9.2	0.40

where r is the complete root, s is the stele (peeled root) and f_s is the fraction of root mass contributed by the stele.

6.3 Results

6.3.1 Soil and the soil exchangeable pool

In this fieldsite, soil Ca concentrations were typically around 10 g kg^{-1} and the average Ca isotopic composition was 0.44±0.07‰ (n=49) [Hindshaw et al., 2011a]. The range of soil Ca isotopic compositions was 0.28 to 0.57‰ and this is the range indicated in the figures. The soil exchangeable (plant-available) pool of Ca comprised 1-5% of the total soil Ca. The Ca isotopic composition of this pool was heterogeneous ($\delta^{44/42}$Ca = 0.28 to 0.52‰, Table 6.2), but not significantly different to the range of bulk soil values [Hindshaw et al., 2011a]. The average $\delta^{44/42}$Ca value of the soil exchangeable pool was identical to that measured in the rhizosphere soil sample and this value (0.40‰) is used as the soil $\delta^{44/42}$Ca value in the calculation of plant-soil Ca isotopic differences. The $\delta^{44/42}$Ca value of the rhizosphere soil was not significantly different from the bulk or organic soil samples of the same soil age (Table 6.3). Root samples had lower Sr/Ca and Ba/Ca ratios compared to the soil exchangeable pool (Fig. 6.1a) indicating that if the soil exchangeable pool would be the only source of cations, then root uptake processes alter these ratios.

6.3.2 Whole Plant Analyses

The Ca concentrations of *Rumex* tissues increased from root (9.34 g kg^{-1}) to stem (10.8 g kg^{-1}) to leaf (31.4 g kg^{-1}). A similar increase in concentration was observed for the two *Rhododendron* specimens analysed, but the absolute concentrations were lower, for example the highest leaf Ca concentration was only 9.19 g kg^{-1} (Table 6.3). The Ca concentration of the leaves of both species was more than three times that found in the whole root. Flower tissue had higher Ca concentrations than stem tissue in *Rhododendron*, but in *Rumex*, flower tissue had lower Ca concentrations than both the stem and the root. Within the root, the cortex (outer root) had higher Ca concentrations than the stele (inner

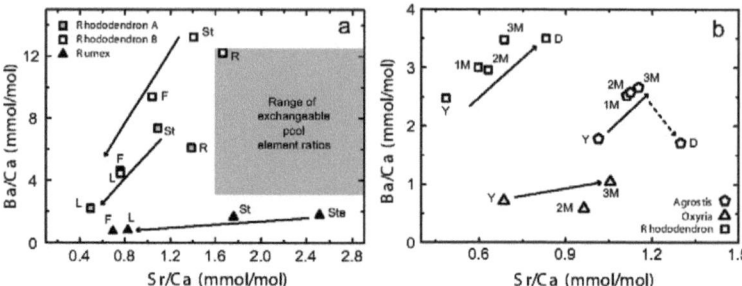

Figure 6.1: Change in Ba/Ca and Sr/Ca ratios (a) along the transpiration stream and (b) over the growing season in the leaves. Arrows highlight the direction of change. In (b) the dotted arrow points to a straw (dead leaf) sample which may be influenced by leaching processes and decomposition on the ground. R=root, St=stem, F=flower, L=leaf, Ste=stele, Y=young leaves, 1M, 2M and 3M=1, 2 and 3 month old leaves respectively, D=dead leaves.

root) for both species. In *Rumex*, the cortex was seven times more concentrated in Ca than the stele, whereas in *Rhododendron* it was only twice as concentrated. The alkali earth element ratios Sr/Ca and Ba/Ca decreased along the transpiration stream (stele to leaf) in *Rumex* and the same general trend was observed in *Rhododendron* (stem to leaf) (Fig. 6.1a and Table 6.3).

In both species there was significant Ca isotope fractionation (>0.36‰) between the exchangeable soil pool ($\delta^{44/42}$Ca = 0.40‰) and the roots. Within the *Rhododendron* specimen RhB there was little isotope fractionation from root ($\delta^{44/42}$Ca = 0.04‰) to stem (0.09‰) to leaf (0.00‰), the same was observed for RhA but the tissues were isotopically lighter than those of RhB (Fig. 6.2). The flowers of both specimens were 0.2‰ lighter than the leaves. For *Rumex*, the $\delta^{44/42}$Ca value of root and stem tissue was nearly identical to that of *Rhododendron* but the leaves and flowers had higher $\delta^{44/42}$Ca values ($\delta^{44/42}$Ca = 0.40‰), in the range of soil $\delta^{44/42}$Ca values (Fig. 6.2). The cortex of *Rumex* roots was 0.25‰ lighter than the stele. As the complete root system of a *Rhododendron* specimen could not be obtained, the fraction of root mass in the stele is not known. However, given the measured isotopic compositions and concentrations of the stele and the whole root, it was calculated using equation 6.2, that the mass fraction of stele could change from 10 to 90% and the isotopic composition of the cortex ($\delta^{44/42}$Ca = -0.02‰) would not differ, within error, from that of the stele (0.01-0.05‰).

Whole plant Ca isotopic compositions, calculated using equation 6.1, were 0.23 to 0.54‰ lighter than soil (Fig. 6.2), unambiguously demonstrating that the uptake of Ca from soils by plants is accompanied by an enrichment of light Ca isotopes. In addition to calculating the whole plant Ca isotopic composition, equation 6.1 was used to calculate the isotopic composition of the whole plant at each step of the transpiration stream. The results (Table 6.4) confirm the negligible within-plant fractionation in the *Rhododendron*

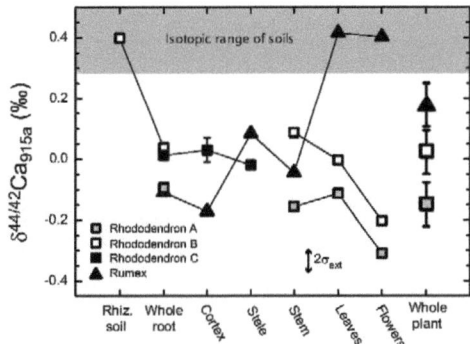

Figure 6.2: Variation of $\delta^{44/42}Ca$ between different plant tissues for two specimens of *Rhododendron* and one specimen of *Rumex*. Rhiz. soil stands for rhizosphere soil - the soil in direct contact with the plant roots. The Ca isotopic composition of the cortex is calculated using equation 6.2. The range in *Rhododendron* cortex $\delta^{44/42}Ca$ is the range obtained when the mass fraction of the cortex is varied from 10 to 90%. The error in whole plant isotopic compositions is 0.14‰. For *Rhododendron*, the whole plant isotopic compositions are calculated assuming the roots contribute 25% to the total mass of the plant.

specimens, but highlight the increase in Ca isotope ratios along the transpiration stream in *Rumex*.

6.3.3 Seasonal variation

The Ca concentration of leaves increased with age, with 3 month old leaves up to six times as concentrated in Ca as young leaves (Fig. 6.3, Table 6.5). The increase in leaf Ca concentration with age has previously been observed in deciduous trees [Guha and Mitchell, 1966] and is due to the lack of retranslocation of Ca, resulting in steady accumulation with time, together with the retranslocation of other nutrients and water loss from older leaves. The observed increase in concentration was most pronounced in *Rhododendron* leaves and least in *Agrostis* leaves (Fig. 6.3). There was an increase in the Ca concentration of *Agrostis* flowers with age (1.3 to 2.3 g kg^{-1}).

Overall the Sr/Ca and the Ba/Ca ratios in the leaves (and flowers) increased with age for all species analysed (Fig. 6.1b). The only exception to this general trend was the Ba/Ca ratio of the dead *Agrostis* leaves which was lower than the 3 month old leaf sample and may have been influenced by leaching and decomposition processes on the ground.

Leaf samples from *Rhododendron* had a constant Ca isotopic composition for the first two months ($\delta^{44/42}Ca = -0.20$‰), then became progressively enriched in the heavier isotopes as the leaves aged (Fig. 6.3, Table 6.5). In contrast, there was no clear change in the leaf Ca isotopic compositions of *Agrostis* and *Oxyria* as the leaves aged. Both *Rhododendron* and *Agrostis* leaves were always isotopically lighter than average soil (0.44±0.07‰) whereas *Oxyria* leaves had an indistinguishable Ca isotopic composition

Table 6.3: Ca concentrations, $\delta^{44/42}$Ca and element ratios for different plant tissues and soils

Sample	Total weight (g)	[Ca] (g/kg)	Sr/Ca (mmol/mol)	Ba/Ca (mmol/mol)	$\delta^{44/42}$Ca $2\sigma_{ext}$=0.07‰	$\triangle^{44/42}Ca_{plant-soil}$[1] $2\sigma_{SD}$=0.10‰
Rhododendron A						
Root	9.53[2]	1.59	1.39	6.09	-0.09	-0.49
Stem	20.7	2.02	1.09	7.35	-0.16	-0.56
Leaf	5.51	9.29	0.49	2.20	-0.11	-0.51
Flower	2.39	3.63	0.76	4.61	-0.31	-0.71
Rhododendron B						
Bulk Soil	-	10.6	7.23	17.4	0.38	
Organic Soil	-	12.2	6.44	13.1	0.38	
Rhizosphere Soil	-	8.03	7.52	28.2	0.40	
Root	2.79[2]	1.16	1.66	12.2	0.04	-0.36
Stem	5.36	2.52	1.41	13.2	0.09	-0.31
Leaf	2.59	7.14	0.76	4.43	0.00	-0.40
Flower	0.42	3.03	1.04	9.40	-0.20	-0.60
Rhododendron C						
Whole Root	-	1.05	1.65	15.7	0.01	-0.39
Cortex[3]	-	*1.45*			*0.03*	-0.37
Stele	-	0.65	1.81	16.6	-0.02	-0.42
Rumex						
Whole Root	2.53	9.34	2.16	1.34	-0.11	-0.51
Cortex	*0.80*	*22.4*	*2.05*	*0.48*	*-0.17*	-0.57
Stele	1.74	3.33	2.51	1.80	0.08	-0.32
Stem	1.99	10.8	1.76	1.69	-0.05	-0.45
Leaf	1.35	31.4	0.83	0.83	0.41	+0.01
Flower	0.82	6.43	0.69	0.76	0.40	0.00

values in italics are calculated from mass balance equations
[1] Relative to rhizosphere soil $\delta^{44/42}$Ca = 0.40‰
[2] Assumes roots contribute 25% to total plant mass
[3] Assumes the cortex and stele contribute 50% each to the root mass

Table 6.4: Whole plant $\delta^{44/42}$Ca compositions (‰) relative to soil ($\triangle^{44/42}Ca_{plant-soil}$) based on mass balance

Fraction	RhA	RhB	Rumex
Complete plant	-0.54	-0.37	-0.23
Ste + St + L + F	-	-	-0.14
St + L + F	-0.55	-0.37	-0.13
L + F	-0.54	-0.42	0.01

Ste = stele, St = stem, L = leaf, F = flower

to soil. No increase in Ca isotopic composition with age was observed in the flowers of *Agrostis* and there was a constant Ca isotopic difference between the leaves and flowers of ~ 0.15‰.

6.3.4 Variation along the chronosequence

Three species were sampled at various points along the chronosequence: *L. alpina*, *Salix h.* and *Salix r.* (Table 6.6). Calcium concentrations of *L. alpina* tended to decrease with soil age, whereas those of the two *Salix* species tended to increase with soil age (Fig. 6.4a). For all species there was no correlation between the Ca isotopic composition of the above-ground plant parts and soil age (Fig. 6.4b). The degree of mycorrhizal infection may be expected to influence the Ca isotopic composition of the plant. There was a general increase of root colonization with soil age across both species (Table 6.6), but

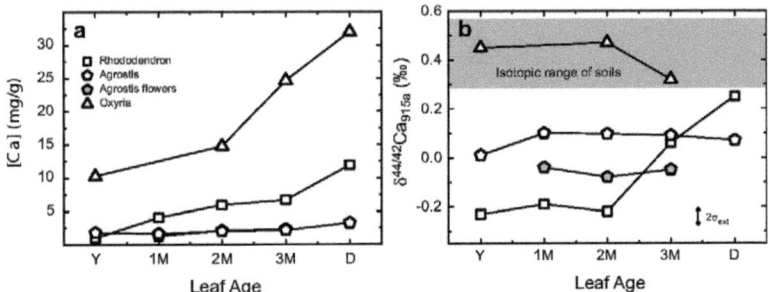

Figure 6.3: Variation of calcium concentration (a) and $\delta^{44/42}$Ca (b) in leaves and flowers (*Agrostis* only) versus tissue age. The range of $\delta^{44/42}$Ca values in the soils is indicated by the grey area. For leaf age, Y denotes young leaves sampled in June; 1M, 2M and 3M denote leaves sampled 1 2 and 3 months after young leaves were sampled; and D denotes dead leaves which for *Rhododendron* were still attached to the plant and for *Agrostis* were collected as straw.

Table 6.5: Ca concentrations, $\delta^{44/42}$Ca and element ratios for samples collected at various times throughout the season

Sample	[Ca] (g/kg)	Sr/Ca (mmol/mol)	Ba/Ca	$\delta^{44/42}$Ca $2\sigma_{ext}=0.07‰$
Rhododendron ferrugineum Leaves				
Young	0.99	0.48	2.47	-0.23
1 month	4.07	0.60	3.01	-0.19
2 months	5.93	0.69	3.48	-0.22
3 months	6.67	0.63	2.96	0.06
Dead	11.9	0.83	3.51	0.25
Oxyria digyna Leaves				
Young	10.2	0.96	0.58	0.45
2 months	14.7	0.69	0.71	0.47
3 months	24.7	1.06	1.04	0.32
Agrostis Leaves				
Young	1.81	1.01	1.78	0.01
1 month	1.65	1.15	2.67	0.10
2 months	1.99	1.11	2.52	0.10
3 months	2.19	1.12	2.58	0.09
Dead	3.24	1.30	1.71	0.07
Agrostis Flowers				
1 month	1.30	0.90	2.29	-0.04
2 months	2.09	1.25	2.85	-0.08
3 months	2.32	1.16	3.03	-0.05

within each species there was a wide range of Ca isotopic compositions which were not correlated with the percentage of roots infected with mycorrhizal fungi. However, one sample which was not infected had a significantly higher $\delta^{44/42}$Ca value than all of the infected samples (Fig. 6.5). Across all species with above-ground biomass data, Ca concentration was correlated with Ca isotopic composition ($r^2 = 0.52$, p<0.01 across all species, Fig. 6.6) with lower $\delta^{44/42}$Ca values relative to soil at lower Ca concentrations. Sr/Ca and Ba/Ca ratios were neither correlated with soil age, Ca isotopic composition nor Ca concentrations in the above-ground biomass of these plants.

Table 6.6: Ca concentrations, $\delta^{44/42}$Ca and element ratios for plants (above-ground biomass only) along the chronosequence

Sample	Soil Age (years)	[Ca] (g/kg)	Sr/Ca (mmol/mol)	Ba/Ca (mmol/mol)	$\delta^{44/42}$Ca $2\sigma_{ext}=0.07‰$	% of mycorrhizal root colonization
L. alpina	7	11.8	1.42	1.27	0.34	0
	8	12.1	1.80	1.18	0.19	34
	14	9.82	1.45	0.97	0.21	38
	59	7.58	2.45	3.03	0.13	55
	65	7.44	2.63	4.16	0.24	66
	68	6.46	1.57	1.58	0.23	47
	73	8.32	1.65	2.46	0.15	72
	76	6.77	1.55	2.25	0.16	30
Salix h.	7	2.89	0.99	0.42	0.11	28
	67	5.29	2.08	4.12	0.05	22
	73	7.31	1.72	1.38	0.14	47
	78	3.52	1.69	3.13	0.06	23
	79	4.87	1.94	1.73	-0.02	35
Salix r.	59	1.43	3.19	6.28	-0.04	39
	129	4.91	1.71	0.34	0.10	70
Ectomycorrhizal sporocaps						
Laccaria montana		0.61	2.59	7.95	0.21	
Russula exalbicans		0.32	3.44	9.07	0.23	

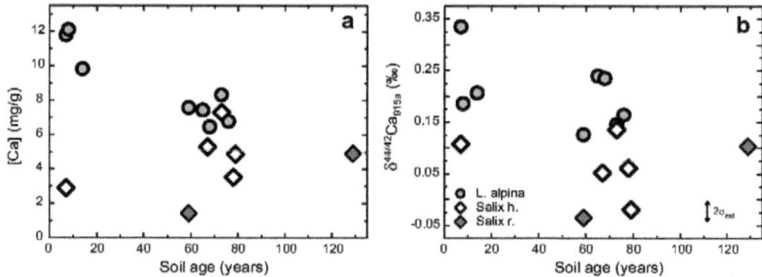

Figure 6.4: Ca concentration (a) and Ca isotopic composition (b) of the above-ground biomass of three plant species versus soil age. Across all three species soil age had no effect on the Ca concentration or the Ca isotopic composition of the plant.

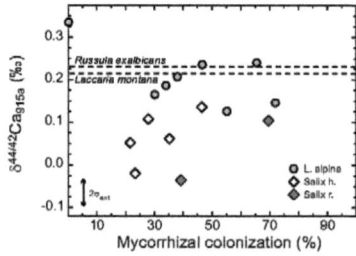

Figure 6.5: $\delta^{44/42}$Ca in above-ground biomass plotted against the percentage of roots infected with mycorrhizal fungi for three different plant species. The measured Ca isotopic composition for the sporocaps of two species of ectomycorrhizal fungi collected from the forefield are indicated by the dotted lines. There is no clear relationship between the percentage of roots infected and above-ground $\delta^{44/42}$Ca values, but a single specimen of L. alpina which has no mycorrhiza is notably heavier than the rest of the specimens.

6.4 Discussion

It is clear from the variation in above-ground Ca isotopic compositions that Ca isotope fractionation is species dependent and that there is also significant intra-species variability in above-ground Ca isotopic compositions (Fig. 6.6). We will first discuss within-plant Ca isotope fractionation with the data from the *Rhododendron* specimens and then discuss inter-species variability.

6.4.1 Calcium uptake processes

Significant Ca isotope fractionation occurs between the soil and the roots of *Rhododendron* (Fig. 6.2), as was also observed for several tree species [Wiegand et al., 2005, Page et al., 2008, Cenki-Tok et al., 2009, Holmden and Bélanger, 2010]. Fractionation is either occurring at the soil-plant interface, or the plant accesses a source of isotopically light Ca.

The exchangeable soil pool could be a potential source of light Ca. However, all the root samples were more than 0.24‰ lighter in $\delta^{44/42}$Ca than the lightest $\delta^{44/42}$Ca value measured in the soil exchangeable pool and therefore additional fractionation must occur between this pool and plant roots. Mycorrhiza act as an additional interface between the soil and the plant, thus the extent of Ca fractionation in a plant might be expected to depend on the percentage of root area infected with mycorrhiza. However, no clear relationship between the extent of root infection and above-ground Ca isotopic composition was found (Fig. 6.5), suggesting that mycorrhiza are not the main control of plant Ca isotopic compositions.

We consider it likely that the isotopic difference between plant roots and soil arises due to a fractionation process occurring during the uptake of Ca into plant roots. Ca uptake is by mass flow, i.e. passive uptake driven by transpiration [Marschner, 1995] and this process is not expected to cause Ca isotope fractionation. Within the root, Ca transport is predominantly apoplastic (between cells), partly due to the toxicity of high levels of cytoplasmic Ca [e.g. McLaughlin and Wimmer, 1999]. The presence of the endodermis, a hydrophobic layer of cells in fully developed root cells, prevents apoplastic Ca transport from the cortex (outer root) to the stele (inner root). Transport across the endodermis is symplastic (within cell) requiring transport proteins [White and Broadley, 2003]. This would be expected to induce measurable isotope fractionation between the cortex and the stele as was recently observed for iron [Kiczka et al., 2010]. The fact that the cortex and the stele of *Rhododendron* had the same isotopic composition strongly supports the view that Ca uptake is primarily apoplastic (between cells) and the majority of uptake must therefore occur at the root apices where the endodermis has not yet developed [e.g. Karley and White, 2009]. Calcium isotope fractionation must therefore occur in the cortex. Studies utilising isotopically labelled solutions have shown biphasic uptake behaviour for divalent cations [Epstein and Leggett, 1954, Drew and Biddulph, 1971] with

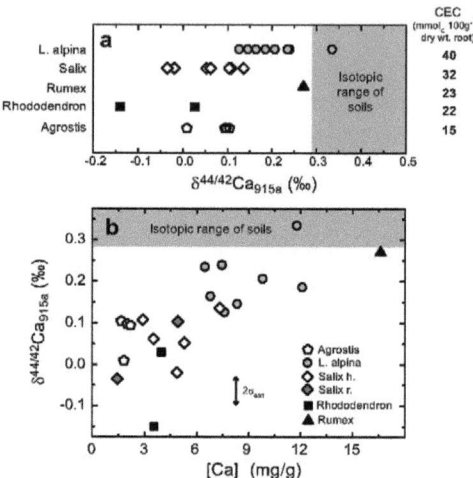

Figure 6.6: (a) Summary of above-ground Ca isotopic compositions of species analysed. The root cation exchange capacity (CEC) for the different plant species at the order level [White and Broadley, 2003] is shown for comparison. Within the dicotyledonous species (all except *Agrostis*) the whole plant Ca isotopic composition increases with increasing root CEC. The Ca isotopic composition of *Rumex* is affected by the precipitation of Ca oxalate in the root cortex and therefore does not follow the trend of increasing $\delta^{44/42}$Ca values with increasing root CEC observed in the other dicotyledonous species. (b) Across all species, there is a correlation ($r^2 = 0.52$, $p<0.01$) between the concentration of Ca (dry weight) and the above-ground Ca isotopic composition. This suggests that root level Ca uptake processes, i.e. the amount of Ca adsorption, determined by root CEC, controls the overall plant Ca isotopic composition.

an adsorbed and an absorbed component. Both adsorption and adsorption processes could fractionation Ca isotopes, resulting in the difference in isotopic composition observed between roots and soil.

The absorbed phase is essentially irreversibly bound Ca which has undergone active transport into cells. Due to low cytoplasmic Ca concentrations, an electrochemical gradient exists from the apoplasm (mmol L^{-1} Ca) into the cell (nmol L^{-1} Ca). Calcium in the cytoplasm is actively removed by pumping it into the vacuole or back out of the cell across the cell membrane. This is achieved by calcium ATPase and Ca^{2+}/H^{+} antiporters [Evans et al., 1991]. It is expected that these active processes will favour the transport of light Ca as a result of incomplete kinetic reactions, resulting in an accumulation of light Ca inside the cell vacuole.

The adsorbed phase is readily exchangeable with the soil solution and it has been shown that the adsorption of magnesium by root cells induces isotope fractionation with the adsorbed phase enriched in heavy Mg isotopes [Bolou-Bi et al., 2010]. Therefore, it is reasonable to assume that Ca isotopes will also fractionate during adsorption. Adsorption is likely to be an equilibrium fractionation process whereby the stronger bonding environment is enriched in the heavier isotopes. The most common adsorption sites for Ca, and other metals, in the apoplasm are carboxyl groups located in the cell walls [Marschner, 1995]. Adsorption studies are not available for the Ca isotope system, but equilibrium fractionation has been invoked to explain the light Ca isotopic composition of carbonate precipitates relative to seawater [Lemarchand et al., 2004, Marriott et al., 2004, Gussone et al., 2006]. An enrichment of light Ca in a carboxyl bonding environment is supported by estimates of bond lengths, which are a proxy for the strength of the bond. During equilibrium fractionation, the heavy isotopes are enriched in the stiffer (shorter) bonds [Schauble, 2004]. Calcium-oxygen bond lengths for hydrated calcium from density functional theory calculations and spectroscopic methods indicate a Ca-O bond length of 2.4 Å [Carugo et al., 1993, Kaufman Katz et al., 1996, Pavlov et al., 1998, Jalilehvand et al., 2001] for Ca in an inner-shell hydration complex with a coordination number (CN) of six to eight. Studies of crystallographic data indicate that bidentate binding of Ca to the carboxyl group results in a Ca-O bond length of 2.5 Å [Einspahr and Bugg, 1981, Carugo et al., 1993] which is in agreement with a ^{43}Ca nuclear magnetic resonance (NMR) study indicating Ca-O bond lengths of 2.5 Å for Ca acetate and Ca oxalate [Wong et al., 2006]. Unidentate binding, which is the most common form of Ca-carboxyl binding [Kaufman Katz et al., 1996], has a shorter bond length (2.4 Å), but due to the oblate shape of carboxylate O, this may not be representative of stronger bonding [Einspahr and Bugg, 1981]. In addition, unlike Ca-carboxyl bonds in a crystal complex, adsorbed Ca may retain part of its hydration shell resulting in an increased CN and longer Ca-O bonds. Based on these results, the adsorbed phase with the weaker, longer bond should be enriched in light Ca. The extent of this fractionation would be dependent on the number of available exchange sites, the functional groups present and the extent of equilibration with the soil solution. Thus, both adsorption and absorption are capable of causing isotope fractionation in the same direction and could be responsible for the light Ca isotopic composition of root tissue compared to soil.

6.4.2 Within-plant fractionation processes

6.4.2.1 Root to leaf and reproductive organs

In the two *Rhododendron* specimens analysed, excluding the flowers, negligible within-plant Ca isotope fractionation along the transpiration stream was observed (Table 6.4). Both adsorption and absorption processes are predicted to fractionate Ca isotopes resulting in an enrichment of light Ca isotopes on adsorption sites on cell walls and inside cells. If absorption of Ca by cells controls the Ca isotopic composition of the apoplasm then the apoplasm would be expected to be isotopically heavier than the contents of root cells. Since Ca in xylem fluid is primarily derived from the apoplasm [Karley and White, 2009], this would result in the stem and leaves being isotopically heavier

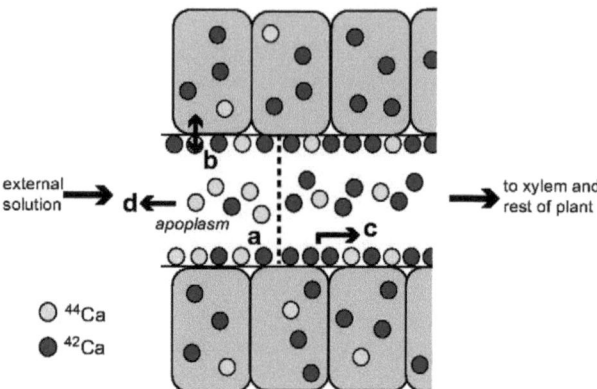

Figure 6.7: Cartoon of the root tip illustrating proposed mechanisms for Ca isotope fractionation during uptake. (a) Adsorption of Ca to cell wall functional groups, preferential binding of ^{42}Ca. (b) Uptake of Ca into cells. The vacuole contains around 75% of the Ca in root tissue and vacuole contents are relatively immobile. Transport across the tonoplast (vacuole membrane) is active [White et al., 1992] and this results in preferential accumulation of ^{42}Ca inside cells. (c) Recently adsorbed Ca is retranslocated to rest of plant, resulting in an enrichment of ^{42}Ca in the xylem fluid compared to the external solution. (d) Only a fraction of the apoplasmic volume can be exchanged with the external solution, represented by ions left of the dotted line, in this way the plant can preferentially expunge ^{44}Ca.

than the root tissue. Since no within-plant Ca isotope fractionation was observed (Table 6.4), adsorption probably controls the Ca isotope ratio of the xylem. In a closed system, adsorption of Ca onto cell wall functional groups would also result in the root apoplasm, and therefore xylem fluid, becoming enriched in heavy Ca isotopes. However, exchange of ions with the external solution readily occurs at the root tips [Marschner, 1995]. In order to explain the inferred preferential adsorption of light Ca and the lack of enrichment of heavy Ca isotopes along the transpiration stream, we suggest the following hypothesis (Fig. 6.7): New cells formed at the root tip provide new exchange sites and Ca fractionates as it binds to these sites. The residual (isotopically heavy) pool of Ca is then removed from the plant by exchange with the external solution. The adsorbed Ca can then be absorbed, remain adsorbed or be retranslocated. Since most of the Ca adsorbed in the root apex is immediately retranslocated [Ferguson and Bollard, 1976], the light Ca isotopes preferentially adsorbed will also be preferentially retranslocated.

The decrease of Ba/Ca and Sr/Ca ratios from stele to stem to leaf in *Rhododendron* (Fig. 6.1a) is consistent with the occurrence of ion exchange during transport through the xylem [Ferguson and Bollard, 1976, Van de Geijn and Petit, 1979]. However, unlike previous studies [Page et al., 2008, Holmden and Bélanger, 2010], no clear increase in $\delta^{44/42}$Ca values from root to leaf, as a result of ion exchange, was observed in *Rhododendron*. This is probably because the length of the xylem 'ion-exchange column' in *Rhododendron* was not long enough, compared to mature trees, to produce significant Ca isotope fractionation, analogous to the explanation proposed by Viers et al. [2007] for the different Zn isotopic compositions of herbs and trees.

The flowers of *Rhododendron* were 0.20‰ lighter in $\delta^{44/42}$Ca in both specimens analysed compared to the bulk plant isotopic composition. The flowers of *Agrostis* were also lighter by 0.15‰ than the leaf+stem $\delta^{44/42}$Ca value. The flowers contain the reproductive organs of the plant and these tissues acquire Ca from the phloem. Ca reaches the phloem by symplastic transport and is thus expected to be isotopically light due to kinetically controlled isotope fractionation during transport of Ca across cell membranes.

6.4.2.2 Seasonal variation in leaf Ca isotopic composition

Significant increases over the growing season were observed in the Ca isotopic compositions and Ca concentrations of *Rhododendron* leaves. Unlike most plant nutrients, Ca is not retranslocated, thus the increase in concentration reflects accumulation over the whole season. As whole leaf Ca isotopic compositions were analysed and since the retranslocation of Ca from leaves is negligible, the change in leaf $\delta^{44/42}$Ca with age must represent a change in the Ca isotopic composition of the xylem fluid, which is the source of Ca to the leaf.

Rhododendron is a perennial woody plant and develops tree rings. Ferguson and Bollard [1976] demonstrated, using a ^{45}Ca tracer, that in autumn, compared to spring, xylem fluid transport was retarded and that this was due to lignin deposition increasing the cation exchange capacity of the xylem. If the Ca isotopic composition were controlled by cation exchange then the Ba/Ca and Sr/Ca ratios should decrease with leaf age as the heavier cations are more strongly retained on the exchange sites. However, a seasonal increase in Ba/Ca and Sr/Ca ratios was observed (Fig. 6.1b). In autumn secondary xylem growth results in smaller cells compared to growth in spring with a corresponding increase in cell wall growth, for which Ca is a major component [McLaughlin and Wimmer, 1999]. We suggest that the relative increased demand for Ca from the stem tissue in autumn (involving active uptake) could therefore lead to an increased $\delta^{44/42}$Ca value of the xylem fluid and the increase in leaf Ba/Ca and Sr/Ca ratios observed (Fig. 6.1b). If correct, this seasonal change in leaf $\delta^{44/42}$Ca values would only be observed in plants with woody stems.

6.4.3 Intercomparison of alpine plant species

Rumex and *Oxyria* are closely related species and both had elevated Ca isotopic compositions ($\delta^{44/42}$Ca = $\sim$0.40‰) in their leaves and flowers. Unlike *Rhododendron*, these plants did not show seasonal variations in leaf Ca isotopic composition, probably because these plants die back in autumn. With respect to the whole plant Ca isotopic composition, the roots were 0.28‰ lighter and the leaves 0.24‰ heavier. These two plants are known to contain high levels of oxalate in order to deter grazers [Landolt and Urbanska, 2003] and Ca oxalate crystals can occur in all parts of the plant [Doaigey, 1991]. Crystals were observed under a microscope in root samples from *Rumex*. Ca oxalate contains a Ca-carboxylate bond and is predicted to be enriched in light Ca isotopes, based on bond-length arguments outlined above (see section 6.4.1). The concentration of Ca in the cortex was extremely high (22 g/kg, cf. 1 g/kg for *Rhododendron*) and this, together with the 0.25‰ increase in $\delta^{44/42}$Ca between the cortex and the stele, strongly suggests that calcium oxalate is precipitating within the cortex. We suggest that the $\delta^{44/42}$Ca of xylem fluid is further increased during passage through the stem due to further oxalate precipitation, resulting in the observed heavy isotopic composition of the leaves and flowers.

Root cation exchange capacity (CEC) is thought to be important in controlling Ca uptake [Haynes, 1980] and because above-ground $\delta^{44/42}$Ca values and Ca concentrations were correlated (Fig. 6.6b), root CEC may also control above-ground Ca isotopic compositions. Excluding *Rumex* because of the additional effect of oxalate precipitation, there is a trend of increasing above-ground Ca isotopic composition and Ca concentration with increased root cation exchange capacity (Fig. 6.6). The greater the number of adsorption sites (higher CEC) the closer the Ca isotopic composition of the plant is to soil (source Ca). *Agrostis* was the only monocotyledonous species analysed and its relatively high above-ground Ca isotope ratio could be due to differences in the composition of the root cell walls [Hose et al., 2001] and demand for Ca between monocotyledonous and dicotyledonous species. Thus, whole plant $\delta^{44/42}$Ca values are largely determined by the root cation exchange capacity which is species specific, but for some species, additional factors are also important, such as the Ca oxalate precipitation observed in *Rumex* and *Oxyria*.

6.4.4 Comparison with other divalent cations

Strontium and calcium are expected to behave similarly during plant uptake and translocation, due to their similar chemical properties. It is also expected that the stable isotopes of these two elements should have similar within-plant fractionation patterns. However, a previous study of stable Sr isotope fractionation within *Rhododendron* plants at the Damma glacier fieldsite [de Souza et al., 2010] concluded that Sr fractionated in the opposite sense to Ca during within-plant fractionation based on a comparison with previously published Ca isotope data on trees [Wiegand et al., 2005, Page et al., 2008]. This study measured *Rhododendron* from the same fieldsite (different specimens) allowing the stable isotope fractionation of Sr and Ca to be directly compared in the same species. Our results show that the discrepancy is solely due to comparison between different species, and directly comparing stable Ca and Sr isotope ratios relative to soil from the same species (*Rhododendron*) shows that the direction of stable isotope fractionation between soil and different tissues is the same (Table 6.7). In *Rhododendron*, both within leaves of different age and between different tissues, there is a strong correlation of Ca and Sr concentrations ($r^2 = 0.97$ and 0.80 respectively), suggesting that these two elements are behaving similarly within the plant. The similarity between Sr and Ca stable isotope fractionation in plants indicates that either the active transport mechanisms cannot distinguish between Ca and Sr and/or that adsorption fractionates both cations similarly as a result of both cations competing for identical adsorption sites [Epstein and Leggett, 1954].

Table 6.7: Comparison of stable Ca and Sr isotope ratios relative to soil in *Rhododendron* (different specimens)

	Sr (‰)[1]		Ca (‰)	
	$\Delta^{88/86}$Sr	95% CL[2]	$\Delta^{44/42}$Ca	$2\sigma_{SD}$
root-soil	-0.29	0.14	-0.34	0.10
	-0.18	0.10	-0.47	0.10
			-0.37	0.10
flower-soil	-0.54	0.14	-0.58	0.10
			-0.69	0.10
stem-soil	-0.24	0.14	-0.29	0.10
			-0.54	0.10

[1] All Sr data from de Souza et al. [2010]
[2] CL stands for confidence limit

Magnesium is expected to behave similarly to Ca during root uptake processes [Ferguson and Bollard, 1976]. It has been suggested that the primary fractionation of Mg isotopes occurs at root level during adsorption of Mg onto cell wall carboxyl groups [Black et al., 2008, Bolou-Bi et al., 2010], similar to Ca. However, these two isotope systems fractionate in the opposite sense to each other: the adsorbed phase, organic molecules and whole plants are enriched in heavy Mg [Black et al., 2007, 2008, Bolou-Bi et al., 2010] and light Ca isotopes. Assuming equilibrium fractionation controls adsorption and binding to organic molecules (which contain similar functional groups to root adsorption sites) then this means that Mg is more strongly bound to organic molecules compared to water and the reverse is true for Ca. Plant physiology reflects this: Mg is vital in the functioning of many enzymes and is a key component of chlorophyll, whereas Ca is used in signalling and must be able to be released quickly from binding sites in response to external stimuli [Marschner, 1995]. In terms of isotope fractionation, Zn behaves similarly to Mg: the roots of plants and organic molecules are enriched in heavy Zn isotopes and leaves are isotopically lighter than roots [Weiss et al., 2005, Viers et al., 2007, Jouvin et al., 2009, Moynier et al., 2009]. The enrichment of heavy Zn isotopes in the adsorbed phase is consistent with inorganic adsorption experiments [Balistrieri et al., 2008, Juillot et al., 2008]. In contrast, although inorganic surfaces and dead (eliminating active cell uptake processes) bacterial surfaces have been shown to be enriched in heavy Cu isotopes [Balistrieri et al., 2008, Navarrete et al., 2011a], living plants show an enrichment of light Cu isotopes in root tissue [Navarrete et al., 2011b].

This difference is probably due to the redox behaviour of Cu. In summary, for non-redox active elements, isotope fractionation patterns in plants may be able to be predicted from knowledge of the fractionation factors between aqueous complexes and common organic ligands (e.g. oxalate).

6.4.5 Effect of plants in the biogeochemical Ca cycle

Plants strongly fractionate Ca compared to soil pools, and would thus be expected to impact soil pool Ca isotope values. No influence of plants on soil, water or porewater $\delta^{44/42}$Ca was detected in this catchment, due to the very low vegetation density [Hindshaw et al., 2011a]. However, the effect of plants on Ca cycling is likely to change as vegetation succession proceeds. Calcium from litterfall can be mobilised and re-used by plants, but its composition is species dependent: old *Rhododendron* leaves had a Ca isotopic composition identical to soil whereas those of *Agrostis* had a lighter isotopic composition. In addition, as vegetation succession proceeds an increasing proportion of Ca is incorporated into biomass, especially into woody tissues. In the late stages of vegetation succession the size of the Ca pool in plants can exceed that of the exchangeable pool [Johnson, 1992]. Thus, it is conceivable that during succession, the Ca isotopic composition of the exchangeable pool will change as a result of vegetation development and this could potentially affect porewater and thus stream water isotopic compositions. However, it is unclear how important vegetation succession versus mineral weathering is in controlling the dissolved fluxes of streams [Gorham et al., 1979, Taylor and Velbel, 1991]. Ultimately the Ca isotopic composition of stream water will depend on the relative size of the respective fluxes. For example, Holmden and Bélanger [2010] found that 80% of the Ca in stream water in a forested watershed in Saskatchewan was biologically derived despite 90% of Ca being internally recycled. Calcium isotope variability in the Strengbach catchment (France) has been attributed to seasonal vegetation control [Schmitt et al., 2003a, Cenki-Tok et al., 2009]. On the other hand, Tipper et al. [2006b, 2008a] concluded that vegetation did not affect the Ca isotopic compositions of Himalayan rivers. These results are not contradictory, Saskatchewan is located on the Precambrian Shield and the mineral weathering rates of shield terrains are generally low. The Himalaya however, are typified by high chemical (and physical) weathering rates. Thus, where mineral weathering rates are low, fractionated reservoirs of Ca due to biological activity may be observed. In such locations, large-scale ecosystem disturbances such as forest fires or clear cutting, which are known to cause high transient Ca fluxes as Ca is released from decaying humic matter [Likens et al., 1970, Balogh-Brunstad et al., 2008], could cause significant changes in soil pool and runoff $\delta^{44/42}$Ca values.

6.5 Conclusions

Several species of plants collected from a recently glaciated, granitic catchment were shown to have a bulk Ca isotopic composition which was isotopically lighter than soil, but the magnitude of $\Delta^{44/42}Ca_{plant-soil}$ was strongly species dependent and was not dependent on soil age. The Ca isotope fractionation patterns observed in *Rhododendron* were explained by two processes: equilibrium fractionation during adsorption of Ca onto carboxyl groups located on cell walls, resulting in an isotopically light adsorbed phase and kinetic fractionation during active uptake by cells resulting in the accumulation of light Ca inside cells. Comparing all species, the higher the Ca concentration of the plant, the smaller the Ca isotope fractionation relative to soil, with high Ca concentrations linked to high root cation exchange capacities. This result indicates that the amount of adsorption of Ca by roots is a key factor in determining the Ca isotopic composition of the plant. The Ca isotopic compositions of tissues from *Rumex* and *Oxyria* were additionally influenced by the precipitation of isotopically light Ca oxalate in the root cortex (and other tissues).

Calcium is an essential plant nutrient and stable Ca isotopes have the potential to refine our understanding of plant uptake and utilisation processes. This study only analysed bulk plant tissues. The next step would be to analyse plant tissues in finer detail e.g. tree rings and xylem fluid and exploit the combination of stable isotopes and radioactive tracers. In addition, Ca isotope fractionation during adsorption remains to be experimentally verified and the form of binding during adsorption investigated. The results from this study indicate that stable Ca isotopes are a promising tool for understanding the biogeochemical Ca cycle in terrestrial environments and elucidating species differences in Ca utilisation by plants.

Chapter 7

Conclusions

The focus of this thesis was the investigation of chemical weathering processes and the fractionation of Ca isotopes in a glaciated, granitic catchment in the Swiss Alps (Damma glacier).

Changes in the chemical composition of stream water can be used to identify water sources and weathering processes. Detailed temporal sampling of stream water chemistry revealed significant and systematic seasonal and diurnal variations in element concentrations, element ratios, $\delta^{18}O$ and $^{87}Sr/^{86}Sr$. The combined use of several tracers and hydrological modelling allowed both water and chemical sources to be identified. Snow and ice melt were the principal water sources and the two main chemical sources were surface melt and the weathering of sub-glacial sediments. The chemical composition of the sub-glacial component was dependent on the residence time of water in the sub-glacial drainage system at the time of sampling. Water residence time affects chemical weathering processes as was evinced by the temporal response of cation/Si ratios. Fast flow paths, which are dominant in summer and during the day, were characterised by high cation/Si ratios. Whereas slow flow paths, which are dominant in winter and at night, were characterised by low cation/Si ratios. The use of additional stable isotope systems such as Li, Si and Mg will help constrain which weathering-related processes are causing the observed changes in cation/Si ratios. The water chemistry at Damma is controlled by glacial melt and the routing of melt water through the glacier. However, the hydrological control of stream water chemical compositions observed in this study is not confined to glaciated catchments. The high weathering rates obtained from glacial catchments are primarily due to their high annual discharge, but a much greater understanding of hydrology will be required in order to fully assess the extent of hydrological control on chemical weathering. For example, little is known about how changes in the annual distribution of precipitation will alter weathering rates. Does one heavy rain storm have the same effect as several smaller rain events? In a glaciated catchment a large proportion of the melt-water does not interact with soils in the pro-glacial zone, whereas in a non-glaciated catchment with the same discharge through rain events, the whole catchment area will be affected. Similarly, changes in precipitation source from snow to rain will affect the timing of the discharge response: a melting snow field provides a sustained input of water long after the initial precipitation event. Thus, it is not so much a question of determining the relationship between annual runoff and weathering rates, but understanding where the water goes, the antecedent hydrological conditions of the catchment and how these parameters affect weathering rates.

A compilation of glaciated and non-glaciated granitic and basaltic catchments permitted an investigation of the effect of changes in climate and erosional regime on weathering rates. Clear differences in the response of basalt and granite to these parameters was observed. Although the weathering rates of both lithologies were related to runoff, changes in temperature and erosional regime only affected the weathering rates of granitic catchments. Catchments which had been subject to physical weathering (frost-shattering, glaciation) had higher cation relative to silica fluxes than catchments with low erosion rates (e.g. shield terrain). The increased sensitivity of granite to changes in climatic parameters and physical erosion could be due to the texture of the rock: granite is coarser grained than basalt. Glaciation, through its changes in hydrology and erosional regime, will have the most pronounced affect on chemical weathering rates if the glaciers form over granite. Even after de-glaciation, these landscapes will retain altered weathering rates due to the fractured rock and glacial flour left behind. Repeated glaciation cycles in the same watershed are likely to enhance differences in weathering compared to ice-free areas, thus the evolutionary history of a catchment must be considered when comparing modern-day weathering rates.

Calcium isotopes were measured in the main Ca pools, and in the inputs and outputs of Ca to the forefield area, in order to investigate if and where fractionation occurs during biogeochemical Ca cycling. This site was particularly suited to the investigation of Ca isotope fractionation during initial weathering processes due to the homogeneous lithology and sparse vegetation cover. The Ca isotopic composition of soil and stream water was indistinguishable from that of rock, indicating that the dissolution of granite does not strongly fractionate Ca isotopes. In this small area (10 km^2), rock and soil Ca isotopic compositions were very heterogeneous. This finding underscores the importance of detailed field sampling in order to gain a representative viewpoint of the system. Although no Ca isotope fractionation was observed at the bulk scale, a sequential extraction procedure of bulk soil revealed that the exchangeable soil Ca pool had a different Ca isotopic composition from bulk soil. In addition, although the average seasonal Ca isotopic composition of stream water was the same as rock, one sample taken at the peak of snow-melt had a significantly higher Ca isotope ratio. These signs of Ca isotope fractionation, if confirmed, are unlikely to be caused by vegetation due to the sparse vegetation cover present at Damma. I suggest that these fractionation processes arise as the result of the preferential sorption (including ion exchange) of light Ca onto surfaces (soil or sediment particles). In the stream, this process occurs in isolated pockets of water which form during winter, and in spring the water which is flushed out is enriched in heavy Ca isotopes. In the soil exchangeable pool, the magnitude of Ca isotope fractionation may be dependent on the type of functional group present, resulting in the different $\triangle^{44/42}Ca_{exchangeable-bulksoil}$ values observed between different soils.

The field site at which these samples were collected has only been ice-free for the last 150 years and the biogeochemical Ca cycle is dominated by mineral weathering. In order to gain a more global view of Ca isotope fractionation in natural systems a wider variety of environments should be sampled. Clearly, an important step would be the investigation of Ca isotope fractionation during sorption to a range of organic and inorganic substrates in controlled laboratory experiments.

The uptake and utilisation of Ca by plants induces significant fractionation of Ca isotopes. The primary uptake step is the adsorption of Ca onto cell walls at the root tip, primarily onto carboxyl groups. In order to explain the light Ca isotopic composition of the root tissue, light Ca isotopes must be preferentially adsorbed. Within-plant fractionation of Ca was shown to be strongly species dependent. In woody plants, there was a seasonal increase in the leaf Ca isotopic composition, probably as a result of changes in secondary xylem growth at the end of the growing season. In herbs and grasses, the plants die back at the end of each season and no secondary growth occurs, thus no seasonal change in leaf Ca isotopic compositions was observed in these species. Internal Ca isotope fractionation (root to leaf) is observed in tree species, but was not observed in the woody species in this catchment, probably due to the small stature of these plants. In two species of the *Polygonaceae* family, which contain Ca oxalate to deter grazers, the effect of Ca oxalate precipitation was observed in the Ca isotopic composition of the plant tissues. The direction of fractionation strongly suggests that Ca in a carboxylate bonding environment is enriched in the light isotope, consistent with the proposed enrichment of light Ca on cell wall adsorption sites. The Ca content of each species was strongly determined by the species-specific root cation exchange capacity which in turn determines the overall plant Ca isotopic composition. The higher the plant Ca concentration, the closer the whole-plant Ca isotopic composition was to soil. This indicates that root level adsorption processes have a major influence on whole plant Ca isotopic compositions. Such root level adsorption processes likely control the root and whole-plant isotopic compositions of other non-redox active metal cations such as Sr, Mg and Zn. The pronounced species differences in Ca isotope fractionation make Ca isotopes a promising tool for constraining Ca utilisation processes within plants. Sampling at smaller scales, for example stem segments, xylem fluid, tree rings, would reveal more information on the fractionation processes occurring. Once the main fractionation processes have been identified the effect of vegetation on soil and water Ca isotopic compositions will be able to be more accurately accounted for.

Chemical weathering is an inherently complex process, and the processes occurring at the Damma glacier catchment are only representative for a small fraction of the Earth's surface. However, it is only by conducting detailed inter-disciplinary studies at a small catchment scale, in conjunction with inter-catchment comparisons, that the factors controlling weathering rates and the associated inter- and intra-catchment variability will be elucidated. An improved understanding of these factors will result in a greater insight into the role of this major geochemical process in modulating Earth's landscapes and climate.

Appendix A

Methods

A.1 Sample Preparation

Sample preparation and analysis was conducted in the clean laboratory of the Isotope Geochemistry group and in the laboratory space of the Soil Chemistry group (both at ETH Zurich). All acids for sample preparation (HNO_3, HCl and HF) were single distilled before use by in-house infrared distillation of analytical reagent grade acids. The concentration of all new batches of acid were checked by titration with 1M NaOH using a few drops of phenolphthalein solution (1% in ethanol) as the indicator. Acid dilutions were performed volumetrically. Acids were stored in 1 L teflon bottles, but some dilute acids were stored in HDPE bottles. Acids for cleaning beakers were not distilled before use. Bottles for storing acids were cleaned using the acid which they were going to store, with the exception of storage container for HF, which was cleaned using HNO_3. Teflon bottles were heated during cleaning and HDPE bottles were not. Super-Q ring system water ($\sim$18 MΩ cm) is hereafter referred to as 'MQ water' and Milli-Q Element Ultapure water ($\sim$18.2 MΩ cm) as 'MQe water'.

Teflon beakers (Savillex, 7 and 15 mL) were used for column chemistry work and sample dissolutions. Teflon beakers used in column chemistry work were cleaned by half filling the beaker with $cHNO_3$ (the 'c' denotes that undiluted concentrated acids were used), closing the beaker and leaving on a hotplate at 120 °C for several days. The beakers were then rinsed copiously with MQ water and placed in a pyrex beaker of water which was heated for 4 hours. The beakers were then dried inside a fume hood. For beakers which had been used to digest samples, a few drops of cHF were added to the first step (addition of $cHNO_3$). After, the HNO_3/HF mixture was discarded and a few mL of HCl was added to the beaker and again left on the hotplate for several days. The cleaning procedure was then identical to that previously described.

Polypropylene centrifuge tubes (various sizes) were used to store samples. The tubes, including caps, were cleaned in bath of cold 5% HNO_3 for several days. The tubes were then rinsed copiously with MQ water and left to dry inside a fume hood. Pipette tips were first cleaned in custom made cleaning racks by soaking in 3M HNO_3 at a mild temperature for 4 hours, they were then rinsed with MQ water and then soaked in warm MQ water for 4 hours before drying.

Water samples were collected in HDPE bottles (60 mL and 1L). The bottles for cation analysis were cleaned using 2% HNO_3 and those for anion analysis were cleaned using MQ water. The bottles were filled with the respective cleaning reagent and left to stand for several days. The bottles were subsequently emptied and rinsed three times with MQ water.

A.1.1 Soils, rocks and mineral separates

At each of the sampling locations a 50x50 cm metal frame was placed on the ground and after the removal of above ground vegetation the soil was sampled at depths of 0-5 cm (samples BLXa) and 5-10 cm (samples BLXb). Roots and stones were removed from each sample before sieving at 0.8 cm. Each sampling point was sampled in triplicate and the soils pooled together to provide one sample for each depth at each site. This composite sample was typically 6 kg. The soil was air-dried at 40°C and sieved at 2 mm before storage.

Soil samples (0.2 mg) were digested using either microwave digestion or by dissolution on a hotplate. Samples for microwave digestion were placed in teflon tubes and a mixture of 5 mL $cHNO_3$, 0.5 mL cHF and 0.5 mL cHCl was added. After the microwave digestion, samples were carefully transferred to teflon beakers and water was added to the teflon tubes to help remove samples where necessary. The teflon tubes were cleaned in 7M HNO_3 overnight, rinsed with copious amounts of MQ water and left to dry in an oven. For samples digested on a hotplate, 8 mL $cHNO_3$, 1 mL cHCl and 1 mL cHF was added to a teflon beaker containing the soil sample. The beaker was sealed and left on a hotplate at 120°C for several days. All samples were dried down at 70°C to prevent fluoride formation and redissolved in 2-5 mL $cHNO_3$. Samples were repeatedly dried down and redissolved until the solution was clear. Where necessary an ultrasonic bath was used to aid dissolution. For 'stubborn' samples the composition of the added acid mixture was adjusted. Finally, samples were dried down and redissolved in 10 mL 2% HNO_3 and transferred to 15 mL centrifuge tubes.

Rock samples and mineral separates were prepared as described by de Souza et al. [2010] and Kiczka-Cyriac [2010]. Briefly, rock samples were crushed using a jaw crusher and ground to a fine powder using a rotary disc mill. Mineral separates of biotite, plagioclase and K-feldspar were obtained by use of heavy liquids, magnetic separation and hand-picking. Mineral separates and rock samples were digested identically to the soil samples.

A.1.2 Water samples

Water at each sampling site was collected in a bucket which was rinsed with river water before filling. Temperature and pH were recorded in situ (Hanna HI 98160 pH meter). All water samples were filtered in the field (with the exception of snow samples which were filtered once melted) using a filtration unit with a hand pump and 0.2 μm nylon filters. The top part of the filtration unit was cleaned with river water and the first 100 mL of filtered water was used to clean the bottom part of the filtration unit and rinse the sample bottles. After filtration, samples for cation analysis were acidified to pH 2 with cHNO$_3$. All water samples were transferred to a fridge for storage within 12 hours for seasonal samples and within 30 hours for diurnal samples. Used filter papers were stored but were never analysed for the > 0.2 μm fraction.

Rain water samples were collected using a 1.5 m high plastic cylinder located near the meteorology station. The cylinder had a plastic funnel at the top feeding into a 2 L plastic bottle and the top of the funnel was protected by a fine mesh to prevent insects from flying in. This bottle was emptied every two weeks. Although this method predominantly collected wet precipitation, a small amount of dry deposition may also have been collected if it was wind-blown into the funnel and subsequently dissolved. Snow samples were packed into pre-cleaned HDPE bottles wearing sterile gloves, after removal of the top few centimetres of the snow pack and were allowed to melt naturally. Porewater samples were collected from suction cups which had been installed in 2007 at depths of 5-25 cm and groundwater samples were collected from tube wells which reached the water table. Rain, snow and ground water samples were filtered as described for stream water samples. The suction cups had a built in 0.45 μm filter and pore water samples were not filtered further.

A filtered water sample (50 mL) was titrated with 1.1 mM HCl and alkalinity was calculated from the titration curve using the Gran method [Stumm and Morgan, 1996]. The concentrations of other species were determined by IC, ICP-OES and ICP-MS.

A.1.3 Plant samples

All plant samples were washed with deionized water, dried at 35°C and ground using either a tungsten carbide rotary disc mill or a zirconium oxide mixer mill depending on the sample size. The majority of samples were digested using a mixture of cHNO$_3$ and H$_2$O$_2$ (30%) in a microwave [Kiczka et al., 2010]. The rest of the plant samples were prepared as follows: 15-20 mg of ground plant material was ashed at 550°C in silver crucibles for 15 min. The powder was transferred to a teflon beaker using MQe water. The water was evaporated and 2-3 mL cHNO$_3$ was added, the sample was ultrasonicated for 2-3 hours and then heated at 80°C for 2-3 days. The lid was removed and once the sample had nearly dried down 0.5 mL H$_2$O$_2$ (30%) was added. The sample was dried down and 2 mL cHNO$_3$ was added and then heated at 80°C for 2-3 days. The sample

Table A.1: Calibration standards for anion analysis

Anion	S/5 mg/kg	S/10 mg/kg	S/100 mg/kg	RSD* %
F^-	0.6	0.3	0.03	11
Cl^-	2.0	1.0	0.1	6
Br^-	4.0	2.0	0.2	-
NO_3^-	4.0	2.0	0.2	3
SO_4^{2-}	4.0	2.0	0.2	4
PO_4^{3-}	6.0	3.0	0.3	-

*RSD is the relative standard deviation of repeat sample measurements

was dried down and 2 drops cHF were added and dried down. Then 0.5 mL HNO_3 was added and the sample heated at 120°C for one hour, ultrasonicated for 15 minutes and dried down. Finally, the sample was redissolved in 10 mL 2% HNO_3.

A.2 Analytical procedures

A.2.1 Element concentrations

Anions (Cl^-, F^-, SO_4^{2-}, NO_3^- and PO_4^{3-}) were analysed by ion chromatography (IC, 761 Compact IC, Metrohm AG, Switzerland). A mixture of 3.2 mM Na_2CO_2 and 1.0 mM $NaHCO_3$ was used as the eluent and the column (Metrosep A SUPP 5 150) was rinsed with 50 mM H_2SO_4 and MQ between each sample. The flow rate was 0.7 mL/min and the pressure was kept around 8.2 MPa. Calibration standards were made from a Fulka multielement ion chromatography anion standard solution by volumetric dilution. Three calibration standards were used (Table A.1). Phosphate and bromide in stream water samples was always below the detection limit of the IC (<0.05 mg/kg). During the course of a measuring session, the calibration standards were periodically remeasured as samples to check for instrument drift.

Major and some trace elements were measured by inductively-coupled plasma optical emission spectrometry (ICP-OES) (Vista-MPX, Varian Inc., USA) and trace elements in water samples were measured by inductively-coupled plasma mass spectrometry (ICP-MS) (Elan DRC-e, Perkin Elmer, USA). Samples of water were analysed directly and samples of soil and rock were diluted 10-150 times prior to analysis with 2% HNO_3. The stock calibration standard for ICP-OES contained 20 mg/L of Si, Ca, K and Na, 4 mg/L of Mg and 1 mg/L of 20 other elements including Sr. Calibration solutions (4-6) were made by volumetric dilution from the stock solution. The concentration range of the calibration solutions was determined by the nature of the samples to be measured (i.e. rock or water). The stock calibration standard for ICP-MS measurements contained 53 elements at either 2 or 200 µg/L. A nine point calibration curve was used. The majority of trace elements measured were below the detection limit (10 pmol L^{-1} - 10 nmol L^{-1} depending on the element) in water samples. The concentrations of the detectable trace metals in water

Methods

Table A.2: Accuracy and precision of cation analysis, 1SD errors are indicated

Cation	SLRS-4 μg/kg	ICP-OES μg/kg n=14	ICP-MS μg/kg n=3	FAAS μg/kg n=3	RSD* %
Si					7
Ca	6200	6179±494	-	-	8
Mg	1600	1619±86	-	-	4
Na	2400	2057±95	-	2060±60	12
K	680	652±27	-	631±48	9
Al	54	56±21	42±3	-	7
Ba	12.2	13.6±2.0	12.6±0.3	-	2
Mn	3.37	4.00±2.5	3.14±0.3	-	7
Sr	26.3	27.3±2.4	28.4±0.7	-	2

*RSD is the relative standard deviation of repeat sample measurements

samples are presented in Table B.2. Initially, K and Na were measured by flame atomic absorption spectrometry (FAAS). However, since the accuracy of measurements was comparable with ICP-OES, this technique was not used further. To monitor the accuracy of cation measurements the water standard SLRS-4 (National Research Council Canada) was repeatedly measured during each measuring session (Table A.2).

Calculated charge balance errors (CBE) for water samples were less than 2.5%, indicating the accuracy of the anion and cation measurements. Some higher charge balance errors (up to 20%) were caused in winter samples by measuring the alkalinity back in the laboratory over 12 hours after sample collection due to adverse weather conditions in the field. This could be due to oversaturation with respect to CO_2: degassing occurred, altering the carbonate equilibrium in the water sample, leading to an overestimation of $[HCO_3^-]$ in the sample. This effect was also noted by Yde et al. [2005] and highlights the importance of immediate measurement in the field.

Element concentrations of the rocks and soils were also determined by X-ray fluorescence spectrometry (XRF) (Spectro X-Lab 2000, Spectro Analytical Instruments, Germany). Four grams of finely ground sample were mixed with 0.9 g wax powder with the aid of 2 mixing balls for 8 minutes. The mixing balls were removed and the powder was pressed into a pellet using a hydraulic compactor (Specac) with a pressure of 15 tons. The pellet was then placed in the XRF machine to be measured. The concentrations of Ca in soil determined by XRF were always within 20%, and mostly within 10% of the values determined by ICP-OES.

A.2.2 Soil sequential extraction procedure

Sequential extractions were performed on the 0-5 cm bulk soil samples. As the upper soil layers are more influenced by organic matter, this fraction was considered more likely to exhibit fractionation between pools than the deeper 5-10 cm layer. A procedure based on Nezat et al. [2007] and Rauret et al. [2000] was used and is summarised in Table A.3. Initially, a four stage procedure was used but a fifth step was added since aqua regia was

Table A.3: Sequential extraction procedure for 0.5 g of soil.

Step	Target Pool	Reagent	Volume (ml)	T (°C)	Reaction time
1	Exchangeable	1 M NH_4Cl	20	25	16 h
2	Organically bound	H_2O_2/HNO_3	2x2.5	85	2 x 1 h
		1 M CH_3COONH_4	15	25	16 h
3	Phyllosilicates	1 M HNO_3	10	180	3 h
4	Residual 1	Aqua Regia	10	120	1 week
5	Residual 2	$HNO_3/HCl/HF$	10	*	*

* microwave digestion

not able to dissolve all of the residual material. Three soils were analysed spanning the complete age range (Soils BL4a, BL17a and BL24a, see Fig. 2.3).

For each soil the procedure was carried out in triplicate (Soil BL17a) or duplicate (Soil BL4a and BL24a) to cover sample heterogeneity and the results were then averaged. After each step the solution was centrifuged, the supernatant stored and the solid residue was carried forward to the next step. During Step 2 hydrogen peroxide was added in small amounts to avoid a violent reaction and the total volume was minimised before adding CH_3COONH_4. In the next two steps (Table A.3) the sample was heated with no shaking and in Step 5 a microwave digestion (3:1:1 mixture of HNO_3, HCl and HF) was used. Each supernatant collected was filtered at 0.2 μm using a hand-held filter holder (Swinnex-47, Millipore, US) with a syringe. The filter holder was flushed with MQ to remove the dead volume. The filtered sample was dried down before being diluted in 2% HNO_3 for elemental analysis by ICP-OES. A procedural blank was also analysed and found to be negligible. Due to incomplete dissolution of target pools, sequential extraction procedures may themselves induce fractionation and phases in addition to the target phase will inevitably be attacked [Wiederhold et al., 2007]. Unintended isotope fractionation may have occurred, but it is assumed that any artificial fractionation effects caused by this procedure were negligible.

A.3 Purification of samples for isotope analysis

A.3.1 Calcium

Due to potential isobaric interferences (e.g. from double charged Sr) and ionisation inhibition on a TIMS filament, each sample must be purified by ion-exchange chromatography prior to isotope analysis. Preliminary tests with 1 M and 1.4 M HNO_3 as elutant acids did not give satisfactory separation of Mg and Mn from Ca. Previous methods developed for relatively pure samples [Skulan et al., 1997, Schmitt et al., 2003a, DePaolo, 2004, Kasemann et al., 2008] gave good separation of Ca from Mg and other alkali and alkali earth elements but were found to be inadequate for soil and rock samples, due to the higher concentrations of Al and Fe, which partially elute with Ca when using 1-2 M HCl (Fig. A.1). Fe and Al impurities have previously been shown to reduce the Ca ion yield and thus beam stability in TIMS measurements [Boulyga, 2010]. It is interesting to note that in 1 M HCl, Al and Ca have theoretical distribution coefficients of 60.8 and 42.29 respectively [Strelow, 1960] i.e. Al should elute after Ca. This did not occur and there is in fact a second Al peak before Ca. This could be due to a matrix effect since loading pure Al onto the column results in the 'correct' elution scheme.

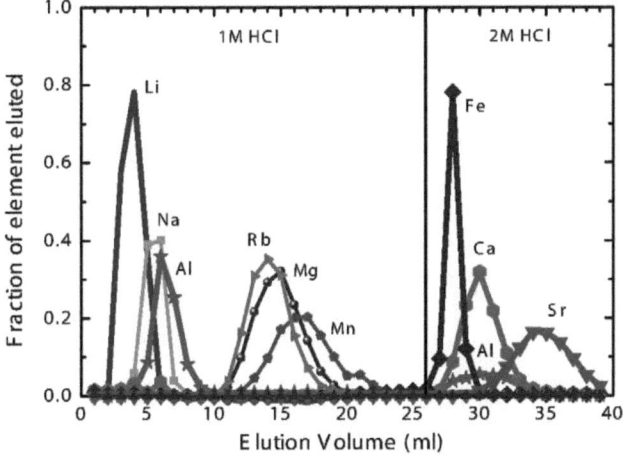

Figure A.1: A typical elution curve obtained for a basaltic rock using 1.6 ml of AG50W-X8 resin (200-400 mesh)

A four column procedure was thus developed with specific columns for the removal of Sr, Fe, Al and other cations. For each sample a volume of liquid corresponding to 3-4 μg of Ca was dried down together with 22-30 μL of a 7 ppm ^{43}Ca-^{46}Ca spike which was added prior to processing the sample through column chemistry (Table A.4) so that any fractionation caused by the less than 100% yield was corrected for. Each collected fraction was dried down and redissolved in the conditioning acid of the next column for loading. The final collected fraction was split in half: one half was dried down, dissolved in 2 ml 1M HNO_3 and transferred to a centrifuge tube for storage and the second half was dried down, redissolved in a few drops of concentrated HNO_3 and dried down again in preparation for loading onto filaments. The AG50W-X8 and the AG1-X4 resins were washed repeatedly before use with MQe and the resin was not changed between each sample. The Sr-spec resin cleaning is outlined in section A.3.2. When not in use the resin was stored in the columns in MQe. The overall yield of whole procedure was estimated to be 95%. The procedural blank through chemical separation was less than 10 ng which is over 300 times smaller than the smallest sample processed. Sr and Fe were separated using Biorad Micro bio-spin columns with an internal diameter of 8 mm, a resin height of 10 mm and a 1 mL reservoir. The last two columns were Biorad Poly-prep columns with an internal diameter of 8 mm, a resin height of 32 mm and a 10 mL reservoir.

1. Remove Sr and Ba
~0.5 mL of Eichrom Sr-Spec resin

Procedure	Reagent	Volume (mL)
Clean	3M HNO$_3$	1
	MQe	1
	6M HCl	1
	MQe	1
Condition	3M HNO$_3$	1
Load	3M HNO$_3$	0.2
Elute	3M HNO$_3$	4x0.2
Clean	3M HNO$_3$	1
(elute Sr and Ba)	MQe	2x1
	6M HCl	1
	3M HNO$_3$	1
	MQe	1

2. Remove Fe
~0.5 mL of AG1-X4 resin (100-200 mesh)

Procedure	Reagent	Volume (mL)
Clean	2M HCl	0.5
	6M HCl	1
	2M HCl	0.2
	MQe	2x1
	2M HCl	0.2
Condition	6M HCl	1
Load	6M HCl	0.2
Elute	6M HCl	5x0.2
	6M HCl	1
Clean (elute Fe)	MQe	4x1
	2M HCl	0.2
	6M HCl	2x1
	2M HCl	0.2
	MQe	1

3. Remove Al
~1.6 mL AG50W-X8 resin (200-400 mesh)

Procedure	Reagent	Volume (mL)
Clean	2M HCl	3
	6M HCl	10
	2M HCl	2
Condition	1M HNO$_3$	2
Load	1M HNO$_3$	0.2
Elute Al	0.1M HF/1M HNO$_3$	2x1
	0.1M HF/1M HNO$_3$	5
	0.1M HF/1M HNO$_3$	3x1
	6M HNO$_3$	7
Clean	6M HCl	5
	conc HCl	5
	6M HCl	5
	2M HCl	2
	MQe	5

4. Remove alkali and alkali earths
~1.6 mL AG50W-X8 resin (200-400 mesh)

Procedure	Reagent	Volume (mL)
Clean	2M HCl	3
	6M HCl	10
	2M HCl	2
Condition	1M HCl	2
Load	1M HCl	0.2
Elute alkali and	1M HCl	2x1
alkali earths	1M HCl	2x10
	1M HCl	4x1
	2M HCl	9
Clean	6M HCl	5
	conc HCl	5
	6M HCl	5
	2M HCl	2
	MQe	5

Table A.4: Full Ca separation procedure, including column cleaning steps. Fractions marked in red contain Ca and were collected.

A.3.2 Strontium

The separation procedure for Sr (Table A.5 and Fig. A.2) was based on the procedure developed by de Souza et al. [2010] and Deniel and Pin [2001]. Teflon columns were used which had an internal diameter of 4 mm, a resin height of 12 mm and a 10 mL reservoir. The yield of Sr was estimated to be >95%. A batch of Sr-spec resin was pre-cleaned before use by washing the resin with MQe, centrifuging the mixture and discarding the supernatant. After three washes of water the resin was soaked in 6M HCl, after this step the resin was washed with water until the water had a neutral pH. This stock of resin was stored in MQe. After each use the resin was discarded and the columns (with frits) were stored in MQ water. Periodically, the columns (with the frits removed) were soaked in warm cHNO$_3$ to remove the build up of sticky residues. The column chemistry procedure differed slightly between samples which were to be measured by TIMS and those which

Table A.5: Column chemistry procedure for the separation of Sr from matrix elements. ~150 µl of Eichrom Sr-Spec resin in teflon columns.

Procedure	Reagent	Volume (mL)
Clean	MQe	2
	6M HCl	2
	MQe	10
	3M HNO$_3$	2
	MQe	2
Condition	3M HNO$_3$	2
Load	3M HNO$_3$	0.05
Elute	3M HNO$_3$	4x0.1
Elute Ba	7M HNO$_3$	1.2
	3M HNO$_3$	2x0.1
Collect Sr	MQe	1.5

were to be measured by MC-ICP-MS. Samples for MC-ICP-MS were obtained from a pre-cut of the column chemistry procedure to purify Mg for Mg isotope analysis of the water samples (separation performed by Tipper see e.g. Tipper et al. [2008b]). For TIMS work, a volume of water containing 1 µg Sr was obtained. All samples were dried down and re-dissolved in 50 µL 3M HNO$_3$ for loading. At the end of the procedure, the final fraction for TIMS samples was split: one half was dried down and redissolved in a few drops of cHNO$_3$, dried down again, redissolved in 2 mL 1M HNO$_3$ for storage, and the second half was dried to a small drop after the addition of cHNO$_3$, this was a precaution against losing the sample as a result of static. Samples for MC-ICP-MS were dried after the addition of cHNO$_3$ and dissolved in 1 mL 2% HNO$_3$.

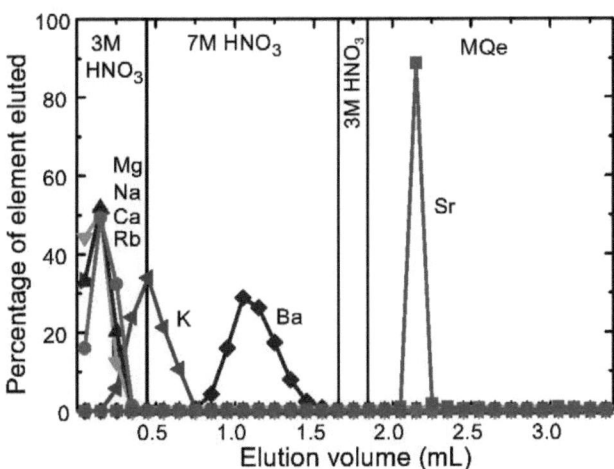

Figure A.2: A typical elution curve obtained for a synthetic rock standard using ~150 µl of Eichrom Sr-Spec resin.

A.4 Analysis of Ca isotopes

All samples were measured by thermal ionisation mass spectrometry (TIMS) (Triton, Thermo Fisher Scientific). Two microlitres of 1M HNO_3 was added to solid product of the column chemistry and half of this volume i.e. 0.7-1 μg of Ca was mixed with 1 μL of Ta phosphate activator solution [Birck and Allègre, 1978] and loaded onto degassed rhenium filaments. Filaments were degassed at 4 A for 2 h using a commercial degassing unit (Thermo Fisher Scientific) with modified electronics and later by an in-house built unit based on a design by Spectromat. Filaments were degassed at least a day in advance of use as liquid did not 'stick' to freshly degassed filaments. Loading capillaries were cleaned in 2% HNO_3 followed by MQe and were flushed before use 1M HNO_3 followed by MQe. The filament was heated with a current of 0.8 A and the solution was added a drop at a time, with the next drop added just before the previous one had completely evaporated. Once all the solution had been added the current was increased until the filament started glowing red. Parafilm was not used during loading as no benefit was observed using it and it removed the hazard of bending the filament. The activator solution was taken from a large drop on a piece of parafilm and throughout the loading session the volume of activator solution used had to be decreased, presumably due to evaporation, to prevent fluffy deposits forming on the filament which were easily knocked off. On the other hand, if too little activator solution was used then the deposits became too 'flaky'. Samples were run as single filaments in static multicollector configuration utilising the full spread of cups (L4 - ^{40}Ca, L2 - ^{41}K, L1 - ^{42}Ca, H1 - ^{43}Ca, H2 - ^{44}Ca, H4 - ^{46}Ca). ^{41}K was monitored in order to correct for any isobaric interference of ^{40}K on ^{40}Ca.

Filaments were heated until the smallest beam was greater than 10^{-12} A (typically ^{42}Ca) to minimise baseline errors. Filaments were heated quickly to start with (250 mA/min) up to 2500 mA, thereafter the heating rate was decreased to 100 mA/min up to 3000 mA, 50 mA/min up to 3600 mA and then 25 mA/min. During the slow stage of heating the beam was focussed, the peaks centered and, if necessary, the zoom optics were adjusted. The heating stage typically lasted 30-45 minutes and the final filament current was typically 3600-3800 mA. Beam stability was highly variable between samples and sometimes adjustment of the current during the measurement run was required to either prevent the ^{40}Ca beam exceeding 5×10^{-10} A or to prevent the beam from 'dying'. It is possible that this beam behaviour, including rare unpredictable jumps in intensity, was caused by heterogeneous evaporation behaviour of the loaded sample [Fantle and Bullen, 2009]. The presence of K suppressed the Ca beam, but once all the K had evaporated the Ca beam would rapidly increase (measurement runs were not commenced whilst K was being detected). Data blocks which showed a clear drift in ratios, which was nearly always attributed to dying beams, were discarded. Amplifier rotation was employed to cancel out differences in amplifier gains and the baseline was recorded before each run.

Methods

A run consisted of 28 blocks of 20 cycles with an integration time of 8 seconds. This was later reduced to 14 blocks when it was observed that the external reproducibility was not affected.

Filament holders, roofs and sides were reused. Holders and roofs were cleaned with an abrasive drill and then all parts were cleaned in MQ followed by a 3% RBS solution in an ultrasonic bath. All parts were then rinsed with MQ and dried in an oven at 80°C.

In order to correct for instrumental mass bias, a ^{43}Ca-^{46}Ca double spike was used. The optimal sample to spike ratio was calculated to be 20 [Rudge et al., 2009]. This spike pair was chosen since ^{44}Ca and ^{42}Ca are used for normalisation in the calculation of radiogenic anomalies [Russell and Papanastassiou, 1978] and ^{48}Ca cannot be measured at the same time as ^{40}Ca due to the limitation of mass dispersion. Double-spike deconvolution was performed offline following Siebert et al. [2001] and Rudge et al. [2009].

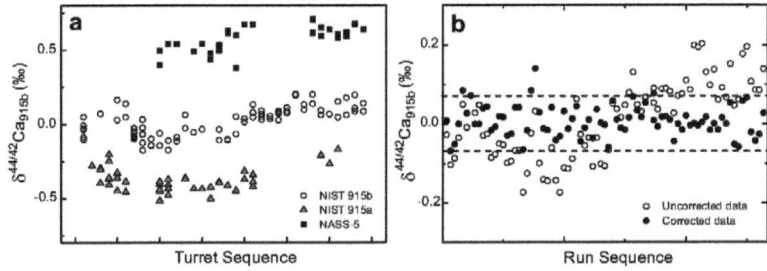

Figure A.3: (a): Long term uncorrected measurements of the three main standards used in this study: 915a, 915b and NASS-5. This illustrates the problem of drift between turrets. **(b):** The result of the correction applied to 915b measurements. The correction essentially brackets the samples on each turret using 2 or more standards (see main text for more details). The dashed lines depict the $2\sigma_{ext}$ error of 0.07‰.

Session to session drifts of standard measurements have been reported on TIMS instruments [Krabbenhöft et al., 2009, Simon et al., 2009]. I also observed this drift in repeated measurements of SRM 915b over time (Fig. A.3) but corrected for it by 'bracketing' each group of samples on one turret with at least two standards. Due to the limited supply of SRM 915a, SRM 915b was used as the primary standard. For each turret, the measured SRM 915b values were averaged and corrected to zero, this same correction was then applied to all samples run on the same turret. This correction varied from -0.20 to +0.14‰. Values are then expressed in standard delta notation:

$$\delta^{44/42}Ca = 1000 \left\{ \frac{\left(\frac{44\,Ca}{42\,Ca}\right)_{sample}}{\left(\frac{44\,Ca}{42\,Ca}\right)_{SRM915b}} - 1 \right\} \quad (A.1)$$

Repeat analyses of SRM 915b gave $\delta^{44/42}$Ca= 0.01±0.01 (n=79, 95% confidence level). For all samples, the $2\sigma_{ext}$ of the standard (0.07‰) is used as a measure of the external reproducibility. Finally, samples were converted to a delta value with respect to SRM 915a (+0.35‰) as the use of SRM 915a as the primary standard follows the recommendation by Eisenhauer et al. [2004]. The $\delta^{44/42}$Ca ratio is used, since unlike $\delta^{44/40}$Ca it is not affected by potential radiogenic enrichments. Although the wide absolute range in measured standard values is reduced by normalising the average of the SRM 915b measurements on each turret to 0 (Fig. A.3), the resultant range is still large and thus although samples run in duplicate on the same turret generally reproduce very well, caution should be exercised in drawing conclusions over small differences in samples which have only been measured two or three times.

A.4.1 Radiogenic Ca

The average ϵ^{40}Ca for SRM 915b was -0.5±2.8, 0.3±2.3 for 915a and 0.1±3.7 for seawater (all errors $2\sigma_{SD}$). These latter two values are in agreement with the values published by Amini et al. [2009] and Simon et al. [2009] for those standards.

A.4.2 Inter-laboratory comparison of Ca isotopic standard measurements

Calcium isotope data have been reported in a variety of notations. In order to compare data the following conversions are used: $\delta^{44/40}$Ca$_{915a}$ = $\delta^{44/40}$Ca$_{seawater}$ - 1.88 [Hippler et al., 2003] and $\delta^{44/42}$Ca = 0.476.$\delta^{44/40}$Ca (assuming equilibrium fractionation [Sime et al., 2005]). The conversion factor in the latter equation is different from previous authors who have either used 0.488 (assuming kinetic fractionation [Holmden, 2009]) or 0.500 (approximation [Hippler et al., 2003]). This wide choice in correction values could lead to problems in data comparison in the future as precision improves. The measured value of the new NIST standard SRM 915b agrees very well with the published value of Heuser and Eisenhauer [2008], but is lower than the value reported by Reynard et al. [2010]. For the bone meal standard SRM 1486 there is a discrepancy between the value I measured and that of Heuser and Eisenhauer [2008] of 0.1‰. I also measured a larger $\delta^{44/42}$Ca value for BHVO-2 compared to Amini et al. [2009]. However, my value for another basalt standard, BCR 1, is in good agreement with Simon and DePaolo [2010] and that of BCR 2 reported by Amini et al. [2009] and Wombacher et al. [2009] (assuming both BCR standards are isotopically the same as is the case for Mg [Tipper et al., 2008b]). Differences may arise because rock standards and the bone meal standard have more complex matrices in comparison to SRM 915a and SRM 915b which are essentially pure carbonates. Although the data compilation (Table EA-3) contains values from at least three different seawater standards (IAPSO, NASS-5 and Mediterranean Seawater) it can be assumed that modern sea water is isotopically homogeneous [Zhu and Macdougall, 1998, De La Rocha and DePaolo, 2000, Schmitt et al., 2001]. The wide range in reported seawater values (spanning over 0.2‰) could be due to differing chemical purification

Table A.6: Ca isotopic standards and reference materials reported in $\delta^{44/42}$Ca relative to SRM 915b (Errors are 2 σ_{SEM})

Study	915b	SW	SRM 1486	BHVO-2	BCR 1	BCR 2
This study	0.35±0.02 (n=46)[1]	0.95±0.03 (n=30)	-0.37±0.02 (n=8)	0.55±0.03 (n=4)	0.44±0.01 (n=3)	0.41±0.03 (n=7)
Amini et al. [2009]	-	0.88±0.02 (n=136)	-	0.40±0.03 (n=21)	-	-
Böhm et al. [2006]	-	0.86±0.05 (n=45)[3]	-	-	-	-
Chang et al. [2004]	-	0.92±0.13 (2σ)	-	-	-	-
De La Rocha and DePaolo [2000]	-	0.86±0.04	-	-	-	-
Fantle and DePaolo [2007]	-	0.97±0.04 (n=2)	-	-	-	-
Farkaš et al. [2007]	-	0.90±0.06 (n=25)	-	-	-	-
Gopalan et al. [2007]	-	0.87±0.06[2]	-	-	-	-
Gopalan et al. [2006]	-	0.90±0.02 (n=25)	-	-	-	-
Gussone et al. [2003]	-	0.93	-	-	-	-
Heuser and Eisenhauer [2008]	0.34±0.02 (n=56)	-	-0.48±0.01 (n=142)	-	-	-
Hippler et al. [2003]	-	0.94±0.07	-	-	-	-
Holmden [2009]	-	0.88±0.02 (n=13)[2]	-	-	-	-
Jacobson and Holmden [2008]	-	0.89±0.04[2]	-	-	-	-
Kasemann et al. [2008]	-	0.93±0.03 (n=3)	-	-	-	-
Page et al. [2008]	-	0.96±0.04 (n=15)[2]	-	-	-	-
Reynard et al. [2010]	0.43±0.03 (n=38)	0.96±0.04 (n=13)	-	-	-	-
Schmitt et al. [2003b]	-	0.90±0.04 (n=2)[2]	-	-	-	-
Schmitt et al. [2009]	-	0.87±0.03 (n=6)	-	-	-	-
Sime et al. [2005]	-	0.98±0.08 (n=6)	-	-	-	-
Simon and DePaolo [2010]	-	-	-	-	0.40±0.02 (n=2)	-
Steuber and Buhl [2006]	-	0.90±0.02 (n=34)[2]	-	-	-	-
Tipper et al. [2006a]	-	1.09±0.07 (n=1)	-	-	-	-
Tipper et al. [2008b]	-	0.93±0.03	-	-	-	-
Wieser et al. [2004]	-	0.88±0.11 (n=54)[2]	-	-	-	-
Wombacher et al. [2009]	-	0.89±0.07 (n=5)	-	-	-	0.44±0.10 (n=4)

[1] this represents measurements of 915a relative to 915b.
[2] this represents measurements of 915a relative to seawater.
[3] average of IAPSO and Mediterranean seawater

techniques on a complex matrix or different acid concentrations of solutions for plasma work [Fietzke et al., 2004].

There has been much discussion in recent years as to the cause of the poor external reproducibility of Ca isotopic measurements by double spike on TIMS [Fletcher et al., 1997, Holmden, 2005, Fantle and Bullen, 2009] and how it could be improved. This has focussed primarily on the cup configuration and instrumental settings. Whilst undoubtedly important, laboratories using a wide range of double spikes and cup configurations are still reporting similar levels of reproducibility.

In order to assess the role of chemical purification in affecting reproducibility, it is suggested that a rock standard should also be routinely analysed and reported. The effect of changing the method of chemical purification was clearly shown in Schmitt et al. [2009], where two different methods often yielded quite different results for the same samples.

A.5 Analysis of Sr isotopes

A.5.1 MC-ICP-MS

Solutions (50 ppb) were introduced to the multicollector ICP-MS (Nu Plasma, Nu Instruments, UK) using a PFA microconcentric nebuliser with an uptake rate of ~ 50 μL/min and the aerosol was dried using a desolvator (either Aridus II, Apex Q or Apex HF depending on the session). This gave a minimum ^{88}Sr ion beam intensity of 5×10^{-11} A. Krypton and Rb interferences were corrected using an on-peak zero and an iterative correction for any residual Kr and Rb above background [de Souza et al., 2010]. The mass bias calculated for Sr was assumed to be valid for Rb and Kr. This is valid since the magnitude of this correction was small compared to the external reproducibility. Samples were bracketed by NBS 987 and all samples and standards were normalised to ^{86}Sr/^{88}Sr = 0.1194. Due to a clear drift in the standard values a secondary correction to give NBS 987 ^{87}Sr/^{86}Sr ratios of 0.710250 was used. Each sample was measured at least twice. Reproducibility ($2\sigma_{SD}$) of standards over the course of this study were 0.709182$\pm$66 and 0.708665$\pm$59 for seawater and an in-house standard respectively. Sample standard deviation ($2\sigma_{SD}$) was less than 80 ppm.

A.5.2 TIMS

Approximately 250 ng of Sr was loaded in nitric form together with 1 μL of tantalum phosphate activator solution onto degassed single rhenium filaments as described for Ca (section A.4), with the exception that after the filament was heated to glowing a further drop of activator solution was added [Aciego et al., 2009]. Data acquisition was comprised of 200 measurements with a 4 s integration time in static mode. The exponential law was applied to correct for instrument mass fractionation and all ^{87}Sr/^{86}Sr

Methods

ratios were normalised to $^{86}Sr/^{88}Sr = 0.1194$. ^{85}Rb was monitored to correct for rubidium interferences on ^{87}Sr. At least two different standards (of NBS 987, seawater and an in-house standard) were analysed in each session to monitor machine drift and each sample was measured four times. Similar to ICP-MS, a secondary correction was applied to all data. Reproducibility ($2\sigma_{SD}$) of standards over the course of this study were 0.709181 ± 25 and 0.708669 ± 18 for seawater and the in-house standard respectively. Sample standard deviation ($2\sigma_{SD}$) was less than 40 ppm. Sample 20080624A was measured by both TIMS and MC-ICP-MS and yielded identical results of 0.720637 ± 36 (MC-ICP-MS) and 0.720636 ± 25 (TIMS).

A.6 Analysis of O isotopes

The oxygen isotopic composition of water samples was measured using the CO_2 equilibration method: 200 μL samples of water were pipetted into 12 mL septum-capped vials which were subsequently filled with a mixture of 0.3% CO_2 and He. After equilibration at 25°C for at least 18 h the CO_2/He mixture was measured using a Gas Bench II (Thermo Scientific) connected to an isotope ratio mass spectrometer (Delta V plus, Thermo Scientific). Measurements were calibrated with the international standards SMOW, SLAP and GISP. The results are reported in the conventional delta notation with respect to VSMOW and sample standard deviation ($2\sigma_{SD}$) was less than 0.1‰.

Appendix B

Additional figures and datatables

Table B.1: Major species and $\delta^{18}O$ data for seasonal sampling (Sites C and D)

Site	Date (YMD)	Time (CET)	pH	T (°C)	Ca^{2+}	Mg^{2+}	Na^+	K^+	Si	F^-	Cl^-	NO_3^-	SO_4^{2-}	HCO_3^-	Sr^{2+} (nmol L^{-1})	$\delta^{18}O$
					(μmol L^{-1})											
C	20080527	12:15	6.37	0.2	20.2	3.4	8.5	10.8	11.1	4.6	3.9	27.5	8.5		37.7	-17.04
C	20080610	11:40	5.87	4.2	18.4	2.7	7.9	7.2	12.8	5.7	3.8	16.6	5.6	14.1	22.8	-15.66
C	20080624	14:05	6.45	7.5	11.3	1.6	5.4	4.8	5.1	3.9	3.5	10.1	3.7	12.4	13.5	-14.82
C	20080708	13:40	6.41	8.5	11.9	1.7	5.1	5.1	6.7	3.4	2.7	11.5	4.0	10.6	15.5	-14.63
C	20080722	09:30	5.55	5.6	14.6	2.5	7.8	7.8	14.7	5.0	3.0	11.4	5.2	20.2	20.2	-14.60
C	20080805	13:30	6.72	10.5	7.0	2.9	2.9	5.1	7.5	2.2	1.8	5.3	2.1	6.6	9.2	-14.65
C	20080819	14:20	6.39	10.0	7.1	1.0	5.1	5.1	4.1	3.5	4.1	4.5	2.0	12.8	8.3	-14.30
C	20080902	12:50	5.86	7.0	6.7	1.0	2.3	3.5	3.4	1.8	1.2	4.3	1.9	11.0	8.2	-13.84
C	20080916	14:20	6.00	4.6	17.6	3.4	10.7	11.6	25.3	6.8	2.5	13.9	7.8	25.9	30.0	-13.84
C	20080930	13:15	5.56	4.8	19.7	4.5	16.5	16.7	39.1	8.5	2.8	12.9	8.0	43.6	32.6	-14.00
C	20081014	13:30	5.76	5.3	17.3	3.3	10.1	11.0	19.5	6.7	3.0	17.2	7.3	27.7	29.8	-13.95
D	20080527	10:55	6.64	0.2	20.0	3.0	17.2	11.1	9.4	4.4	12.3	27.9	8.4		36.7	-17.37
D	20080610	10:30	6.59	1.6	18.5	3.6	7.4	9.5	10.6	2.9	4.8	17.5	6.4	31.9	32.2	-15.79
D	20080624	10:30	6.8	3.2	10.7	2.0	5.3	8.0	6.2	2.8	3.1	9.8	3.6	17.3	19.5	-15.03
D	20080708	10:30	6.52	2.8	11.5	2.1	5.5	8.0	8.1	2.8	2.4	9.6	4.2	16.6	20.6	-14.21
D	20080722	13:00	5.38	3.3	10.3	2.1	5.0	7.6	8.1	2.8	2.6	8.3	4.0	27.8	20.4	-14.49
D	20080805	10:35	6.84	3.9	8.3	7.5	5.8	10.4	14.9	2.1	3.8	4.9	2.3	8.6	16.6	-14.38
D	20080819	10:45	6.15	3.6	7.8	2.5	4.0	7.6	7.4	1.7	1.8	4.5	2.7	17.4	15.5	-14.38
D	20080902	09:50	5.69	3.1	8.1	1.5	3.1	5.5	5.2	2.3	1.4	5.3	2.9	12.7	16.0	-13.85
D	20080916	11:00	6.21	1.9	21.7	5.1	13.7	17.7	25.2	5.1	3.6	16.9	12.3	41.1	50.5	-13.38
D	20080930	10:25	6.14	1.7	19.3	4.8	14.0	17.6	32.0	6.5	2.5	17.4	8.9	44.4	40.1	-14.17
D	20081014	10:20	5.71	1.9	22.0	4.1	11.8	14.7	18.7	5.4	8.9	24.1	9.2	36.5	48.6	-13.80
D	20081028				19.4	5.7	16.0	20.2	35.8	6.3	2.0	19.5	10.4	49.8	41.4	-13.80

Table B.2: Trace metal concentration data for water samples. All concentrations are in nmol L^{-1}, except Al which is in μmol L^{-1}

Sample	Al	La	Ce	Th	U	Ti	Mn	Fe	Cu	Zn	Y	Mo	Li	Rb	Sr	Cs	Ba	Ga	Sn	Sb	Pb
A-20080513	2.06	0.03	0.07	0.05	1.31	4.91	17.41	22.06	3.06	55.55	0.40	11.00	20.54	23.86	36.62	0.03	8.48	0.57	0.00	0.40	0.19
A-20080527	1.70	0.02	0.05	0.04	1.32	5.94	15.25	21.55	2.94	31.85	0.23	11.60	17.55	17.17	27.84	0.03	5.83	0.40	0.00	0.17	0.66
A-20080610	2.06	0.02	0.05	0.02	1.04	4.72	6.95	16.50	2.58	45.09	0.15	10.15	14.56	12.93	28.31	0.02	4.79	0.30	0.12	0.14	0.12
A-20080624	1.19	0.01	0.03	0.01	0.97	4.85	13.09	17.54	1.91	34.85	0.10	6.93	11.27	9.73	17.78	0.03	3.72	0.22	0.00	0.09	0.09
A-20080708	1.07	0.02	0.03	0.01	0.81	8.47	16.12	26.57	1.82	44.32	0.10	7.48	15.31	9.25	18.07	0.03	4.11	0.32	0.35	0.60	0.14
A-20080722	1.19	0.01	0.02	0.02	0.87	6.92	13.44	26.93	2.40	21.56	0.09	9.39	14.26	10.12	19.16	0.02	3.56	0.25	0.00	0.15	0.08
A-20080805	1.67	0.01	0.03	0.01	1.00	8.67	16.33	24.51	2.93	28.46	0.07	5.41	6.34	7.30	12.29	0.03	3.07	0.20	0.00	0.11	0.08
A-20080819	0.41	0.01	0.02	0.01	0.93	4.65	12.51	20.14	1.98	13.87	0.06	6.50	10.83	7.74	12.72	0.03	2.37	0.17	0.46	0.08	0.19
A-20080902	1.64	0.01	0.02	0.02	0.90	3.17	13.91	17.04	1.02	24.21	0.05	5.46	13.22	7.24	11.67	0.03	2.86	0.19	0.19	0.08	0.16
A-20080916	2.13	0.02	0.02	0.03	1.65	7.66	13.41	13.41	2.78	33.05	0.18	15.98	26.47	18.06	36.13	0.03	6.32	0.43	0.08	0.08	0.07
A-20080930	1.22	0.03	0.09	0.05	0.70	3.17	9.10	30.22	1.04	23.66	0.18	21.24	18.71	18.07	34.00	0.03	5.15	0.39	1.00	0.18	0.12
A-20081014	1.68	0.02	0.04	0.03	0.93	6.30	8.78	17.20	1.18	17.61	0.11	26.39	26.47	18.73	34.70	0.03	5.06	0.37	1.62	0.18	0.20
A-20081028	3.67	0.02	0.09	0.04	1.09	6.34	3.66	14.85	3.49	45.97	0.20	27.26	18.71	21.79	42.70	0.03	6.71	0.46	5.29	0.15	0.12
A-20091116	0.61	0.02	0.02	0.03	0.79	4.65	2.35	11.25	0.00	26.82	0.20	25.48	9.37	23.47	48.19	0.01	6.58	0.45	0.00	0.10	0.06
A-20090408	0.00	0.12	0.15	0.18	2.20	6.14	4.09	12.87	1.43	0.00	0.94	31.78	17.65	21.55	73.69	0.02	10.98	0.66	0.00	0.05	0.08
B-20080527	1.26	0.02	0.06	0.04	1.72	4.09	18.39	19.80	2.52	33.74	0.19	7.00	18.90	16.51	17.10	0.04	5.52	0.33	1.07	0.08	0.16
B-20080610	2.15	0.02	0.05	0.02	1.63	2.90	7.94	15.27	4.59	62.20	0.15	6.97	15.01	10.60	17.99	0.03	3.79	0.23	0.00	0.41	0.83
B-20080624	0.33	0.02	0.04	0.01	1.32	2.74	14.17	15.27	1.46	26.24	0.11	5.02	12.17	8.01	13.41	0.03	3.02	0.20	0.00	0.16	0.14
B-20080708	7.76	0.01	0.02	0.01	1.51	2.85	22.99	15.63	4.19	67.52	0.08	4.70	21.76	7.28	14.38	0.03	3.31	0.22	16.01	0.16	0.10
B-20080722	1.73	0.01	0.03	0.00	1.74	2.67	16.37	18.71	8.56	53.72	0.09	6.42	16.21	7.27	13.70	0.02	3.45	0.26	0.00	0.11	6.39
B-20080805	0.45	0.01	0.01	0.01	2.21	2.85	16.43	17.81	0.86	27.29	0.04	3.14	14.71	4.87	14.38	0.02	2.41	0.18	1.17	0.45	0.12
B-20080819	0.15	0.00	0.01	0.00	1.05	1.48	16.34	13.12	0.92	15.13	0.03	3.83	14.11	6.00	7.74	0.02	1.61	0.12	2.59	0.11	0.09
B-20080902	1.34	0.01	0.02	0.01	1.19	2.51	17.34	17.66	2.67	36.10	0.07	3.24	14.11	6.54	7.62	0.03	2.78	0.19	0.45	0.09	0.07
B-20080916	2.54	0.02	0.05	0.01	1.75	3.14	16.76	10.14	4.25	42.35	0.10	7.51	33.27	8.99	17.63	0.01	3.80	0.22	0.00	0.10	0.12
B-20080930	1.73	0.01	0.05	0.01	0.84	5.38	9.89	12.46	1.22	43.34	0.07	4.94	26.94	10.07	15.15	0.01	3.94	0.22	2.81	0.17	0.22
B-20081014	2.27	0.02	0.04	0.02	1.84	4.24	18.38	15.21	2.38	13.31	0.14	11.25	34.38	14.51	18.96	0.02	3.97	0.25	0.76	0.19	0.08
B-20081028	1.50	0.02	0.01	0.00	0.81	3.21	3.16	9.97	1.91	47.62	0.16	13.00	8.10	25.60	24.32	0.02	5.54	0.34	1.74	0.18	0.15
C-20080527	1.08	0.01	0.02	0.01	0.55	6.63	27.32	22.27	1.52	36.00	0.05	16.08	20.39	15.49	37.73	0.04	6.50	0.44	1.91	0.15	0.05
C-20080610	0.77	0.01	0.04	0.02	1.17	5.84	5.58	15.37	2.75	42.39	0.13	8.56	12.02	13.27	22.75	0.04	4.54	0.32	0.00	0.19	0.08
C-20080624	1.17	0.02	0.03	0.01	1.32	2.97	13.08	14.69	1.66	36.90	0.11	5.69	12.92	9.00	13.51	0.04	2.93	0.17	0.00	0.17	0.18
C-20080708	0.83	0.02	0.03	0.01	1.50	3.55	19.18	14.73	2.23	81.83	0.07	5.83	18.89	8.75	15.48	0.03	3.24	0.19	0.11	0.14	0.09
C-20080722	1.97	0.01	0.02	0.01	1.16	4.09	8.78	15.37	2.26	18.17	0.09	9.21	13.97	11.43	20.17	0.02	3.95	0.22	55.00	0.20	0.91
C-20080805	2.00	0.01	0.07	0.01	2.05	4.09	93.88	292.61	2.68	31.19	0.56	3.78	49.23	19.38	9.17	0.36	6.31	1.08	0.04	0.16	0.07
C-20080819	1.78	0.14	0.37	0.15	1.39	2.95	13.90	17.13	2.86	24.74	0.05	4.26	14.49	7.19	8.33	0.02	2.08	0.14	33.82	0.13	1.14
C-20080902	1.08	0.01	0.02	0.01	0.69	4.54	22.09	9.13	1.64	28.12	0.08	3.73	21.09	7.22	8.20	0.05	2.44	0.17	0.25	0.11	1.00
C-20080916	1.53	0.01	0.04	0.03	1.25	9.57	10.05	18.28	1.27	18.71	0.15	13.23	25.04	16.75	30.04	0.02	5.34	0.38	0.00	0.16	0.10
C-20080930	1.27	0.02	0.05	0.03	1.55	8.10	4.93	19.33	2.53	48.99	0.11	20.65	17.92	19.09	32.64	0.03	6.35	0.43	0.46	0.16	0.14
C-20081014	4.02	0.02	0.06	0.03	0.45	7.44	14.37	14.73	2.51	24.95	0.07	22.84	25.83	18.10	29.82	0.03	5.13	0.42	0.99	0.19	0.12
D-20080527	2.33	0.02	0.06	0.01	1.01	4.29	30.66	16.29	6.60	74.02	0.05	16.23	17.10	15.27	36.73	0.05	7.05	0.45	0.00	0.22	0.14
D-20080610	2.59	0.01	0.02	0.01	1.01	10.45	8.55	23.55	2.16	22.10	0.11	10.81	13.52	11.02	32.20	0.05	4.47	0.37	0.00	0.17	0.21
D-20080624	0.63	0.02	0.04	0.01	0.54	8.38	17.77	22.40	2.11	34.15	0.07	7.29	12.77	9.08	19.54	0.03	3.68	0.20	0.00	0.17	0.98
D-20080708	1.22	0.01	0.03	0.01	0.63	5.44	15.19	19.60	2.28	36.25	0.04	9.28	21.09	7.22	20.58	0.03	3.64	0.25	29.51	0.38	0.14
D-20080722	1.42	0.01	0.02	0.01	0.58	4.78	14.99	17.95	2.08	19.00	0.05	9.89	13.52	8.55	20.39	0.02	3.58	0.22	5.33	0.35	0.43
D-20080805	0.99	0.00	0.02	0.01	1.53	4.78	14.99	17.95	2.08	19.00	0.05	9.89	13.52	8.55	20.39	0.02	3.58	0.22	5.33	0.15	0.07
D-20080805	4.35	0.27	0.95	0.29	1.53	529.73	235.25	870.71	4.96	43.40	1.31	6.07	112.35	41.57	16.58	0.74	18.81	3.23	0.04	0.19	2.70
D-20080819	1.76	0.05	0.13	0.07	0.59	81.06	52.24	141.58	2.87	38.46	0.24	7.70	30.71	12.32	15.52	0.14	4.60	0.59	3.83	0.20	0.53

(Continued on next page)

Table B.2 – Continued

Sample	Al	La	Ce	Th	U	Ti	Mn	Fe	Cu	Zn	Y	Mo	Li	Rb	Sr	Cs	Ba	Ga	Sn	Sb	Pb
D-20080902	1.25	0.01	0.02	0.01	0.68	5.53	22.99	13.12	1.37	13.59	0.03	7.73	19.66	7.43	16.02	0.02	2.59	0.12	1.11	0.15	0.06
D-20080916	1.58	0.02	0.05	0.01	1.11	8.58	7.64	15.19	1.34	15.59	0.09	22.46	26.47	16.40	50.50	0.02	6.19	0.40	1.26	0.17	0.08
D-20080930	1.40	0.01	0.03	0.00	0.62	10.12	5.27	21.41	0.85	21.31	0.05	22.31	18.55	15.56	40.10	0.02	5.34	0.34	0.73	0.15	0.21
D-20081014	2.03	0.02	0.11	0.00	1.03	9.39	7.80	23.74	2.16	29.61	0.08	45.46	30.82	18.98	48.64	0.03	5.83	0.42	0.75	0.14	0.14
D-20081028	3.34	0.01	0.01	0.00	0.25	2.51	2.07	11.02	0.11	5.60	0.02	19.65	13.17	17.14	41.39	0.00	5.43	0.38	0.19	0.13	0.04
E-20080527	5.16	0.01	0.00	0.00	0.98	3.13	32.89	15.73	6.60	65.45	0.09	19.90	22.04	14.58	38.56	0.05	7.03	0.42	0.00	0.21	0.12
E-20080610	4.28	0.02	0.04	0.01	0.92	2.11	8.24	16.36	7.99	51.57	0.12	11.33	15.31	12.38	37.71	0.03	4.69	0.32	0.00	0.19	0.13
E-20080624	2.53	0.01	0.03	0.01	0.76	6.59	16.73	17.11	5.66	63.25	0.08	8.16	11.50	9.34	20.31	0.05	4.02	0.25	0.00	0.15	0.14
E-20080708	1.23	0.01	0.02	0.01	0.67	10.45	17.05	29.68	4.25	68.81	0.06	10.69	13.37	9.41	24.12	0.03	4.24	0.32	43.30	0.39	0.45
E-20080722	1.26	0.01	0.02	0.01	0.65	14.51	18.66	33.66	3.24	16.24	0.07	9.62	11.42	9.25	20.05	0.04	3.78	0.29	0.00	0.15	0.13
E-20080805	6.27	0.04	0.11	0.05	0.75	63.58	46.21	120.40	10.06	49.90	0.19	6.28	18.30	10.53	13.05	0.13	5.23	0.62	0.59	0.13	0.47
E-20080819	3.44	0.02	0.05	0.03	0.64	12.04	19.16	31.48	5.81	81.04	0.11	9.79	20.24	10.06	19.48	0.15	3.95	0.28	4.51	0.21	0.35
E-20080902	2.07	0.01	0.00	0.00	1.09	5.12	21.83	11.11	2.37	21.14	0.05	11.24	19.19	9.15	22.63	0.02	3.55	0.23	2.21	0.14	0.11
E-20080916	1.98	0.01	0.02	0.01	0.82	9.17	8.79	17.69	1.20	14.74	0.08	22.76	24.73	17.24	50.89	0.02	6.49	0.46	9.07	0.19	0.09
E-20080930	1.43	0.01	0.02	0.04	0.76	17.81	36.91	36.91	1.76	35.68	0.10	22.84	16.55	18.18	40.93	0.04	5.75	0.48	1.28	0.16	0.14
E-20081014	1.83	0.01	0.02	0.00	0.58	4.54	5.15	14.58	1.22	27.00	0.06	36.32	24.88	18.18	45.66	0.02	5.42	0.38	1.45	0.16	0.11
E-20081028	1.40	0.01	0.03	0.00	0.50	7.62	4.24	17.87	0.87	24.25	0.05	22.96	14.36	20.69	45.24	0.04	6.75	0.52	0.62	0.18	0.06
R-20080624	4.24	0.01	0.01	0.01	0.04	10.91	20.44	30.48	8.54	414.24	0.04	0.61	3.35	2.30	2.84	0.04	2.92	0.18	19.94	0.40	1.63
R-20080708	2.92	0.04	0.10	0.00	0.03	2.57	12.12	16.60	6.80	607.24	0.10	0.48	20.24	2.23	13.50	0.01	6.33	0.39	32.74	0.68	0.60
R-20080726	2.65	0.03	0.08	0.00	0.10	6.36	20.93	20.56	6.24	408.51	0.10	2.77	0.00	1.89	3.66	0.02	3.59	0.18	23.71	0.62	0.59
R-20080805	5.07	0.02	0.06	0.01	0.10	2.97	20.62	15.27	7.91	182.67	0.06	0.55	0.00	2.66	10.34	0.01	5.14	0.39	0.00	0.53	0.50
R-20080819	2.11	0.03	0.06	0.01	0.04	5.42	15.70	20.15	3.37	421.30	0.05	0.18	0.00	1.90	6.00	0.02	3.40	0.25	33.94	1.07	0.46
R-20080902	2.62	0.03	0.06	0.00	0.05	1.67	20.85	5.88	5.18	162.54	0.09	0.36	1.46	1.56	4.66	0.00	2.87	0.20	34.72	0.34	0.50
R-20080916	7.15	0.03	0.10	0.00	0.03	2.29	37.01	7.37	9.43	406.81	0.10	0.53	0.90	2.80	18.93	0.00	5.94	0.34	26.34	0.26	0.22
R-20080930	8.83	0.13	0.52	0.07	0.05	34.17	22.73	74.25	18.41	286.22	0.30	1.11	8.26	6.28	18.81	0.07	11.35	1.00	748.57	1.66	0.73
S-20080513	3.04	0.01	0.01	0.00	0.03	2.54	1.45	14.88	6.65	1159.98	0.02	0.28	0.00	0.43	1.66	0.00	1.35	0.06	0.00	0.04	0.23
S-20081028	5.01	0.00	0.00	0.00	0.15	1.15	6.22	7.34	4.71	4280.21	0.03	0.47	0.51	0.57	2.94	0.00	1.72	0.05	0.34	0.32	0.80
S-20090116	7.73	0.01	0.02	0.00	0.05	2.70	9.65	9.49	2.85	2621.42	0.03	0.31	1.30	0.61	1.75	0.07	4.25	0.23	1.78	0.05	0.58
S-20090116	5.42	0.01	0.01	0.00	0.07	1.23	5.17	8.10	4.06	772.42	0.03	0.12	1.93	0.57	1.64	0.00	6.96	0.49	14.70	0.24	0.76
S-20090116	9.22	0.01	0.01	0.00	0.07	0.00	9.24	7.72	11.88	627.19	0.01	0.30	0.98	1.21	2.90	0.01	15.85	1.01	0.13	0.39	0.31
S-20090406	13.59	0.01	0.01	0.00	0.03	1.54	13.33	6.70	58.75	423.23	0.07	0.28	2.18	1.86	6.39	0.01	6.45	0.41	1.76	0.46	1.35
S-20090407	0.00	0.00	0.01	0.02	0.03	0.85	2.54	3.09	2.08	2256.48	0.07	0.17	1.15	0.45	1.06	0.01	0.70	0.07	0.36	0.04	0.51
S-20090407	0.00	0.00	0.00	0.01	0.03	1.85	2.14	7.33	0.43	690.88	0.01	0.07	0.71	0.31	0.45	0.01	0.17	0.03	3.93	0.04	0.18
GW1a	7.81	0.08	0.18	0.31	1.63	8.21	80.30	28.03	6.81	2963.00	0.86	14.14	17.13	37.72	34.56	0.05	9.46	0.59	292.91	0.21	5.28
GW2a	26.47	0.27	1.31	0.55	2.58	154.89	2234.05	268.16	27.10	5473.96	0.86	10.67	122.40	98.23	100.06	0.28	42.44	3.33	306.37	1.36	1.15
SC1a	209.33	0.68	2.48	1.14	0.86	133.05	271.90	465.85	19.29	12336.51	4.58	1.30	59.87	34.11	45.64	0.20	30.83	2.16	1.10	0.49	10.43
SC1c	33.94	0.12	0.42	0.25	1.07	9.37	10.32	44.35	1.54	385.51	1.00	6.39	37.15	22.49	30.57	0.03	11.74	0.70	4.06	0.19	0.60
SC1d	33.17	0.10	0.35	0.32	0.94	7.73	7.68	38.53	1.61	327.78	0.94	7.29	34.07	20.64	29.99	0.01	10.97	0.69	2.36	0.17	0.52
SC2d	38.75	0.78	2.35	1.77	1.19	97.27	193.85	649.20	25.15	511.86	5.47	0.25	161.98	46.11	129.70	12.47	56.50	3.88	2.82	0.47	9.72
SC2b	46.30	0.57	2.16	3.07	3.44	59.21	155.69	319.37	23.66	5264.53	5.01	0.25	257.13	49.32	23.17	19.22	40.39	2.64	0.39	0.79	5.95
SC3a	19.52	0.37	1.24	3.02	3.00	36.41	26.53	177.49	6.74	2686.97	2.66	0.55	141.71	26.84	27.16	0.03	15.80	1.06	0.00	0.47	3.97
SC3c	32.08	0.54	1.36	1.70	2.49	22.66	249.59	255.63	15.34	30059.98	4.56	0.41	164.04	32.93	100.26	0.12	23.77	1.45	2.19	1.50	6.14
SC5	1417.53	0.29	0.79	0.67	1.21	19.50	82.86	154.30	40.43	10252.87	2.10	1.16	87.57	40.94	40.09	134.06	29.03	1.84	2.91	0.37	23.91
A-08:00	5.19	0.02	0.05	0.06	1.14	12.40	12.42	27.62	6.63	75.94	0.13	13.17	19.65	14.25	28.21	0.04	5.34	0.41	0.47	0.25	0.22
A-09:00	3.23	0.01	0.03	0.02	1.07	6.23	7.41	16.94	2.83	42.57	0.12	13.30	21.14	13.82	29.53	0.03	4.51	0.35	0.00	0.21	0.11
A-10:00	1.70	0.01	0.03	0.01	1.12	7.29	8.00	19.41	2.28	46.66	0.10	13.02	13.67	13.24	26.92	0.03	4.51	0.33	0.00	0.10	0.13

(Continued on next page)

Table B.2 Continued

Sample	Al	La	Ce	Th	U	Ti	Mn	Fe	Cu	Zn	Y	Mo	Li	Rb	Sr	Cs	Ba	Ga	Sn	Sb	Pb
A-11:00	1.22	0.01	0.03	0.02	0.97	8.90	9.90	22.61	2.07	53.31	0.12	12.31	21.59	13.09	26.53	0.03	4.64	0.31	0.00	0.13	0.10
A-12:00	1.72	0.01	0.03	0.01	1.09	6.03	10.31	18.99	2.53	37.15	0.11	10.65	18.30	11.97	22.74	0.02	3.98	0.23	0.00	0.13	0.13
A-13:00	5.94	0.01	0.04	0.02	1.05	10.12	14.38	26.40	1.12	26.51	0.12	9.69	18.90	11.97	20.54	0.04	4.28	0.31	0.00	0.15	0.16
A-14:00	29.48	0.03	0.09	0.07	1.30	41.81	31.55	82.99	2.06	37.73	0.20	9.10	27.42	13.96	19.53	0.09	5.42	0.50	0.00	0.10	0.30
A-15:00	10.15	0.03	0.07	0.02	1.10	6.63	15.35	20.33	1.85	38.62	0.20	8.75	21.13	10.89	19.01	0.03	3.80	0.27	0.00	0.11	0.14
A-16:00	2.05	0.01	0.04	0.02	1.09	6.50	12.66	21.19	1.09	22.42	0.14	9.10	15.91	10.85	19.65	0.05	3.76	0.35	0.00	0.10	0.12
A-17:00	0.78	0.01	0.02	0.01	0.97	5.08	10.93	16.73	1.15	21.91	0.08	9.37	17.10	10.80	20.73	0.05	4.11	0.31	0.00	0.10	0.07
A-18:00	0.74	0.03	0.07	0.04	1.10	25.09	16.03	45.84	1.17	67.88	0.15	8.36	16.96	10.66	18.53	0.06	4.06	0.32	0.00	0.09	0.20
A-19:00	0.54	0.02	0.03	0.02	1.36	6.83	9.45	19.31	0.62	13.10	0.11	10.46	18.30	12.09	23.54	0.03	4.55	0.33	0.00	0.15	0.10
A-20:00	0.76	0.02	0.03	0.02	1.39	8.41	9.64	22.13	0.58	41.70	0.12	10.74	16.41	12.16	23.84	0.03	4.69	0.32	0.00	0.12	0.11
A-21:00	1.34	0.02	0.04	0.02	1.34	8.90	8.63	22.63	0.77	17.99	0.11	10.55	20.24	12.09	23.76	0.03	4.97	0.37	0.00	0.13	0.14
A-22:00	3.40	0.02	0.04	0.02	1.32	6.56	8.40	19.67	2.52	24.40	0.11	11.60	16.96	12.67	25.44	0.03	5.19	0.40	0.00	0.14	0.16
A-23:00	0.75	0.02	0.04	0.02	1.57	8.24	8.79	22.80	1.03	17.16	0.14	12.11	14.56	13.88	24.39	0.03	6.27	0.50	0.00	0.08	0.11
A-24:00	70.40	0.02	0.04	0.02	1.39	8.48	10.49	23.52	1.26	12.00	0.13	12.61	20.10	14.07	27.38	0.03	4.89	0.35	0.00	0.15	0.13
A-01:00	36.73	0.02	0.04	0.02	1.21	9.13	7.68	25.10	2.50	7.28	0.13	12.92	14.30	13.88	28.44	0.02	4.97	0.37	0.00	0.12	0.10
A-02:00	3.47	0.03	0.06	0.04	1.32	17.97	12.09	40.66	2.50	29.57	0.17	11.72	19.80	13.35	25.07	0.05	5.31	0.43	0.00	0.11	0.22
A-03:00	0.98	0.02	0.03	0.02	1.23	7.19	6.27	21.43	0.96	16.80	0.10	11.37	12.02	12.72	24.79	0.03	4.58	0.28	0.00	0.12	0.11
A-04:00	2.72	0.02	0.05	0.03	1.32	8.42	9.19	21.46	4.60	31.94	0.13	12.61	20.37	14.39	28.27	0.03	5.20	0.37	0.00	0.13	0.30
A-05:00	1.90	0.02	0.04	0.02	1.45	7.91	7.71	24.74	2.21	175.32	0.13	12.77	15.16	14.65	29.07	0.03	5.19	0.38	0.00	0.15	0.15
A-06:00	3.46	0.02	0.03	0.03	1.33	5.74	7.42	20.09	4.00	46.10	0.12	13.12	14.80	14.80	30.02	0.03	5.49	0.41	0.00	0.19	0.16
A-07:00	1.15	0.01	0.03	0.02	1.32	6.61	5.45	20.75	1.72	23.04	0.12	13.24	16.81	14.06	28.82	0.03	5.61	0.41	0.00	0.12	0.09
A-11:30	0.00	0.03	0.06	0.03	1.42	23.57	14.63	38.35	2.72	45.19	0.20	12.92	24.13	16.58	30.72	0.06	6.26	0.46	0.29	0.18	0.20
A-12:30	0.77	0.08	0.18	0.16	2.06	114.83	53.98	184.27	3.19	5.10	0.41	13.49	48.58	23.43	30.74	0.23	10.18	0.97	0.10	0.17	0.81
A-13:30	0.00	0.02	0.03	0.03	1.54	3.61	6.15	10.69	0.55	0.00	0.16	12.83	19.57	15.78	29.17	0.03	4.96	0.28	0.02	0.18	0.07
A-14:30	0.00	0.02	0.03	0.03	0.96	5.18	6.93	12.01	0.28	0.00	0.15	12.65	16.62	15.08	28.06	0.03	4.97	0.38	0.08	0.17	0.09
A-15:30	0.00	0.02	0.04	0.03	1.02	4.18	6.83	9.50	0.35	0.00	0.14	12.24	17.36	15.26	27.10	0.03	4.74	0.25	0.04	0.16	0.07
A-16:30	0.27	0.03	0.08	0.05	1.51	31.50	19.90	52.88	3.35	0.00	0.27	11.47	21.92	16.88	26.67	0.07	6.10	0.54	0.05	0.17	0.24
A-17:30	0.00	0.02	0.03	0.03	0.91	7.94	7.48	13.07	0.26	0.00	0.14	11.62	19.86	13.40	24.77	0.04	4.76	0.29	0.01	0.21	0.10
A-18:30	0.00	0.03	0.08	0.06	1.44	35.48	20.75	62.42	0.92	0.00	0.25	12.57	26.64	17.32	28.00	0.08	6.67	0.49	0.00	0.24	0.27
A-19:30	0.00	0.09	0.22	0.16	2.57	113.87	55.98	188.57	0.86	0.00	0.47	12.16	42.40	23.21	28.91	0.21	12.62	1.09	0.03	0.13	0.81
A-20:30	0.00	0.02	0.04	0.04	1.39	4.72	6.97	12.06	0.23	0.00	0.15	12.73	21.48	15.23	27.61	0.03	5.65	0.34	0.00	0.17	0.21
A-21:30	0.00	0.02	0.03	0.04	1.29	3.68	6.96	9.83	0.08	0.00	0.14	12.36	20.30	14.91	27.53	0.03	5.53	0.29	0.02	0.17	0.08
A-22:30	0.00	0.02	0.03	0.04	1.20	5.48	7.44	11.96	4.79	0.00	0.16	12.72	20.89	15.98	29.13	0.02	6.00	0.37	0.04	0.15	0.11
A-23:30	0.00	0.02	0.05	0.03	1.25	4.45	7.56	12.35	1.83	0.00	0.15	12.85	19.27	14.95	28.02	0.03	5.70	0.34	0.04	0.20	0.11
A-00:30	1.08	0.02	0.03	0.03	0.98	5.37	15.51	8.36	3.74	0.00	0.13	11.58	20.89	14.80	26.65	0.02	4.55	0.29	0.28	0.20	0.10
A-01:30	0.00	0.02	0.03	0.03	0.97	4.37	7.36	9.67	0.20	0.00	0.14	12.40	19.12	15.14	28.33	0.02	4.56	0.26	0.42	0.17	0.07
A-02:30	0.00	0.02	0.04	0.04	0.93	5.22	46.43	10.36	0.00	0.00	0.13	12.13	19.57	15.28	28.76	0.03	4.74	0.28	0.20	0.13	0.09
A-03:20	2.80	0.02	0.03	0.03	1.16	4.99	6.82	9.98	2.31	0.00	0.17	13.03	20.15	15.33	29.31	0.03	5.93	0.34	0.06	0.15	0.11
A-04:30	0.00	0.01	0.02	0.03	1.12	4.72	7.01	9.83	0.19	0.00	0.14	12.55	21.48	15.00	28.74	0.03	5.83	0.29	0.04	0.23	0.08
A-05:30	0.00	0.01	0.03	0.02	0.88	4.14	6.23	8.40	0.56	0.00	0.14	12.53	15.29	15.37	29.06	0.04	5.80	0.33	0.04	0.22	0.08
A-06:30	0.00	0.01	0.02	0.02	0.85	8.59	8.59	13.62	0.44	0.00	0.12	11.99	23.40	14.56	27.81	0.03	10.93	0.69	0.02	1.48	0.08
A-07:30	0.00	0.02	0.03	0.03	0.87	4.11	6.70	9.01	0.00	0.00	0.14	13.43	20.15	15.81	30.55	0.04	6.67	0.35	0.00	0.13	0.09
A-08:30	0.00	0.02	0.05	0.05	0.97	4.83	6.59	11.32	0.43	0.00	0.13	13.00	20.15	15.89	31.02	0.03	8.25	0.48	0.02	0.18	0.06
A-09:30	0.00	0.03	0.07	0.06	1.26	32.42	20.42	58.35	1.44	0.00	0.24	12.52	24.87	16.61	29.11	0.07	9.45	0.63	0.02	0.24	0.29
A-10:30	0.00	0.01	0.03	0.02	0.98	5.22	7.00	10.55	1.52	0.00	0.13	13.35	19.12	15.95	29.63	0.04	7.21	0.34	0.00	0.16	0.08

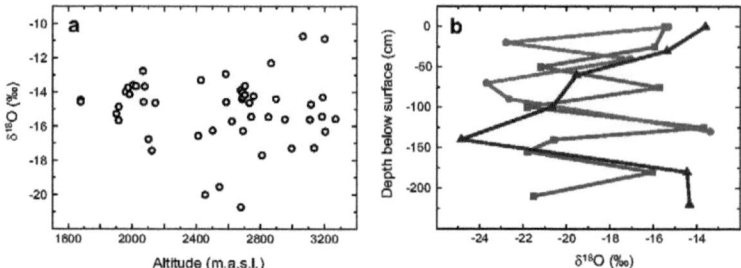

Figure B.1: The variability of $\delta^{18}O$ with altitude (**a**) and depth (**b**). There is no systematic change of $\delta^{18}O$ with either altitude or depth. In (**b**) the colours represent different depth profiles from snow pits; red at 1902m, blue at 2007m and green at 2850m.

Additional figures and datatables

Table B.3: The $\delta^{18}O$ values of snow samples taken from the catchment between 6th and 8th April 2009 to assess variability with depth and altitude. The coordinates refer to the Swiss Grid.

Sample	Easting	Northing	Altitude (masl)	Depth (cm)	$\delta^{18}O$ (‰)
A1	677253	164830	2412	0	-16.57
A2	677176	164801	2429	0	-13.29
A3	677045	164778	2456	0	-20.02
A4	677005	164721	2503	0	-16.27
A5	676934	164687	2546	0	-19.56
A6	676854	164645	2587	0	-14.60
A7	676733	164617	2623	0	-15.73
A8	676667	164566	2680	0	-20.74
A9	676597	164478	2704	0	-14.16
A10	676497	164476	2758	0	-14.25
A11	676393	164525	2811	0	-17.71
A12	676323	164452	2867	0	-12.32
A13	676266	164347	2898	0	-14.41
A14	676141	164326	2954	0	-15.61
A15	676068	164246	2997	0	-17.30
B1	678660	165998	1902	0	-15.30
B2	678660	165998	1902	20	-22.77
B3	678660	165998	1902	40	-17.07
B4	678660	165998	1902	70	-23.68
B5	678660	165998	1902	90	-22.63
B6	678660	165998	1902	130	-13.36
C1	678271	165707	2007	0	-13.60
C2	678271	165707	2007	30	-15.38
C3	678271	165707	2007	60	-19.53
C4	678271	165707	2007	100	-20.67
C5	678271	165707	2007	140	-24.85
C6	678271	165707	2007	180	-14.45
C7	678271	165707	2007	220	-14.33
D1	681081	166798	1679	0	-14.47
D2	681081	166798	1679	0	-14.60
E1	678691	166031	1917	0	-14.88
E2	678675	166009	1917	0	-15.68
E3	678552	165952	1961	0	-13.99
E4	678456	165887	1977	0	-13.74
E5	678387	165805	1985	0	-14.14
E6	678277	165707	2024	0	-13.67
E7	678097	165523	2078	0	-13.68
E8	677966	165353	2101	0	-16.77
E9	677899	165174	2142	0	-14.65
E10	677712	165180	2122	0	-17.43
E11	677972	165384	2072	0	-14.60
E12	678116	165414	2067	0	-12.78
H	676556	165185	2744	0	-15.46
H	676699	165417	2693	0	-16.28
H	676701	165490	2687	0	-14.43
H	676742	165804	2678	0	-13.90
H	676754	165964	2691	0	-14.31
H	676717	166047	2704	0	-13.64
H	676665	166257	2732	0	-14.65
L1	675536	164718	3200	0	-10.89
L2	675640	164874	3135	0	-17.28
L3	675764	166150	3203	0	-16.32
L4	675765	165951	3184	0	-15.44
L5	675828	165640	3109	0	-15.63
L6	675799	165456	3116	0	-14.74
L8	675390	165431	3267	0	-15.59
L9	675560	165309	3189	0	-14.31
L10	675912	165178	3066	0	-10.75
L11	676824	165195	2689	0	-14.02
L12	676958	165009	2585	0	-12.94
S1	676379	165565	2850	0	-15.47
S2	676379	165565	2850	25	-15.94
S3	676379	165565	2850	50	-21.19
S4	676379	165565	2850	75	-15.72
S5	676379	165565	2850	100	-21.80
S6	676379	165565	2850	125	-13.67
S7	676379	165565	2850	140	-20.57
S8	676379	165565	2850	155	-21.79
S9	676379	165565	2850	180	-16.03
S10	676379	165565	2850	210	-21.50

Table B.4: Compilation of annual precipitation corrected element fluxes from granitic and basaltic catchments. B=basalt, Bg=glaciated basalt, G=granite and Gg=glaciated granite. A 'Y' under spot samples indicates that multiple samples from a single river were collected.

Locality	Reference	Lithology	Spot samples	Area km²	Discharge km²/yr	Runoff mm/yr	T °C	Ca*	Mg*	Na*	K*	Si	HCO₃	∑cations
										kmol/km²/yr				meq/m²/yr
Thjórsá	1	Bg	N	7200	12.6	1751	5.2	178	110	607	21	420	1194	1204
Ölfusá	1	Bg	N	5760	13.9	2410	5.1	196	103	561	26	566	1264	1186
Hvítá-S, Gulfoss	1	Bg	N	2000	4.2	2097	4.3	168	90	538	24	489	1117	1079
Tungufljót	1	Bg	N	720	1.5	2073	4.7	104	52	385	19	508	870	716
Sog	1	Bg	N	1050	3.7	3495	4.6	285	133	773	44	664	1868	1654
Hvítá-W, Ferjukot	1	Bg	N	3550	6.0	1681	3.7	121	75	311	14	339	802	717
Hvítá-W, Kljáfoss	1	Bg	N	1685	3.0	1770	2.9	106	50	328	14	351	740	654
Hvítá-S Bruarhlod	2	Bg	N	1987	3.5	1736	9.5	161	92	503	25	346	559	1033
Hvítá-W Kljáfoss	2	Bg	N	970	2.5	2598	5	163	102	707	26	515	898	1262
Vestari Jökulsá	2	Bg	N	815	0.7	810	7.9	92	138	279	23	229	623	764
Austari Jökulsá	2	Bg	N	965	1.1	1119	7.3	58	24	212	10	216	317	388
Skjálfandafljót, Godafoss	2	Bg	N	2428	2.3	964	8	99	69	293	12	209	634	641
Jökulsá á Fjöllum	2	Bg	N	5179	5.2	998	10	95	68	477	15	200	618	820
Jökulsá á Dal	2	Bg	N	3321	4.4	1334	6	116	44	190	7	170	467	517
Jökulsá í Fljótsdal	2	Bg	N	558	1.0	1828	5.5	278	64	217	9	175	808	911
Keldná	2	Bg	N	398	0.7	1784	5.3	64	55	121	6	191	344	367
Geithellnáá	2	Bg	N	184	1.0	5652	7.4	225	159	592	20	517	1062	1380
Skaftafellsá	2	Bg	N	208	1.3	6058	1.7	1505	81	1460	51	586	1542	4685
Hólmsávatn	2	Bg	N	122	3.0	24344	8.7	1715	1116	5970	299	4191	8715	11930
Kuannersuit Kuussuat	3	Bg	N	258	0.5	1900	-3.9	182	82	221	4			752
Stóra Laxá	1	B	N	433	0.6	1479	5	121	67	264	18	417	735	656
Brúará, Dynjandi	1	B	N	670	2.3	3481	4.2	232	81	865	26	847	1729	1516
Brúará, Efstidalur	1	B	N	215	1.3	6047	3.3	300	46	1370	43	1472	1695	2105
Ellidaár	1	B	N	270	0.2	818	4.1	69	25	144	4	162	389	337
Laxá, Vogatunga	1	B	N	142	0.2	1733	5.5	154	74	178	6	243	713	639
Thverá Draghals	1	B	N	44	0.2	3945	4.3	210	140	317	11	552	1287	1028
Grímsá	1	B	N	313	0.7	2178	3.9	158	110	389	12	425	1075	937
Flókadalsá	1	B	N	155	0.3	1690	3.7	133	88	313	11	386	862	765
Thverá, Borgarfjördur	1	B	N	480	0.8	1663	4.3	171	154	278	19	313	1031	948
Nordurá, Stekkur	1	B	N	500	1.1	2222	4.1	175	103	249	11	367	929	816

(Continued on next page)

Additional figures and datatables

Table B.4 – Continued

Locality	Reference	Lithology	Spot samples	Area km²	Discharge km²/yr	Runoff mm/yr	T °C	Ca*	Mg*	Na*	K*	Si	HCO₃	∑cations
										kmol/km²/yr				meq/m²/yr
Sog	2	B	N	439	3.2	7244	7.7	702	401	2544	113	1026	2561	4863
Norðurá Stekkur	2	B	N	391	0.6	1637	7.8	124	94	396	13	245	448	846
Víðidalsá	2	B	N	445	0.2	540	8.7	78	78	163	10	123	344	487
Vatnsdalsá	2	B	N	643	0.3	435	8.9	51	45	163	12	119	322	367
Fnjóská	2	B	N	981	1.3	1305	8	78	44	170	8	225	65	423
Laxá	2	B	N	81	1.2	14321	11	2218	2029	12515	455			21465
Fossá	2	B	N	74	0.3	3378	7.4	102	92	366	9	342	467	764
Ytri-Rangá	2	B	N	245	1.4	5676	8	1413	1247	5537	192	1155		11048
Rivière Langevin	4	B	Y	33	0.1	2470	17.5	267	336	1050	193	1517	2127	2448
Rivière de l'Est	4	B	Y	32	0.2	7060	17.5	755	925	1687	275	1949	4420	5323
R. Grande	5	B	N	18	0.01	543	17	126	122	1326	186	679	2237	2008
R. Salga	5	B	N	10	0.01	734	16	29	20	128	70	282	258	297
R. Faial da Terra	5	B	N	15	0.01	924	15	159	126	197	60	521	869	828
Paraná, Crav 6	6	B	N	50	0.07	1438	25	206	135	120	40	466	750	842
Paraná, Crav 23	6	B	N	13.5	0.01	679	25	145	114	106	29	327	595	653
Kolar	7	B	N			463	25	334	212	268	28	169	1348	1386
Ganjal	7	B	N			463	25	396	197	302	22	183	1504	1511
Machak	7	B	N			463	25	269	572	814	21	256	2544	2516
Kalimachak	7	B	N			463	25	315	403	44	36	387	1407	1516
Khagni	7	B	N			463	25	345	169	89	25	154	1108	1144
Chotatawa	7	B	N			463	25	292	307	256	43	215	1473	1496
Tapti	7	B	N			463	25	340	208	203	27	182	1318	1327
Purna	7	B	N			463	25	233	117	353	47	114	1003	1100
Bori	7	B	N			463	25	379	244	384	48	192	1642	1678
Panjkra	7	B	N			463	25	412	270	671	45	206	1946	2080
Tapti	7	B	N			463	25	295	195	456	38	167	1414	1475
Aranawati	7	B	N			463	25	327	174	299	21	183	1284	1321
Aner	7	B	N			463	25	375	313	297	15	263	1707	1689
Wagurr	7	B	N			463	25	223	98	108	51	112	720	801
Nalganga	7	B	N			463	25	208	68	346	72	87	860	969
Mun	7	B	N			463	25	218	100	205	43	93	706	882
Purna	7	B	N			463	25	143	63	244	36	76	582	693

(Continued on next page)

Table B.4 – Continued

Locality	Reference	Lithology	Spot samples	Area km²	Discharge km³/yr	Runoff mm/yr	T °C	Ca*	Mg*	Na*	K*	Si	HCO₃	Σcations
										kmol/km²/yr				meq/m²/yr
Wardha	7	B	N			463	25	360	239	163	31	204	1376	1394
Jam	7	B	N			463	25	393	175	126	19	161	1017	1281
Kulhakera	7	B	N			463	25	403	155	181	21	129	996	1319
Pench	7	B	N			463	25	392	175	126	19	161	952	1278
Narmada at Garudeshwar	8	B	Y			411	27	215	128	133	9	102	888	828
Mohgaon	8	B	Y			621	25.7	209	133	118	6	200	867	809
Banni	8	B	Y			648	24.1	210	150	196	12	150	1001	929
Patan	8	B	Y			419	26.4	329	157	156		150	1182	1127
Belkheri	8	B	Y			545	25	375	260	231	4	233	1592	1506
Gadarwara	8	B	Y			682	27.8	461	313	312	9	277	1928	1868
Chhidgaon	8	B	Y			586	29.1	454	344	215	7	238	1953	1819
Kogaon	8	B	Y			298	25.9	220	130	111	8	84	915	818
Chandwara	8	B	Y			404	27	214	175	209	4	113	1249	992
Shirdi	9	B	Y			211	26	258	170	142	20	108	590	1017
Nanded	9	B	Y			267	26	219	55	180	24	112	583	750
Krishna at Panchganga	10	B	N			463	24.9	98	74	64	3	135	399	410
Krishna at Panchganga	10	B	N			463	23.8	60	51	42	2	106	359	265
Damma	this study	Gg	Y	10.7	0.03	2856	2.2	36	7	21	24	34	64	131
Leverett	this study	Gg	Y	600	1.08	1800		119	30	138	93	70	388	528
Rhone	11	Gg	Y	26.2	0.08	3100	1.2	42	7	24	26	39	93	148
Mittivakkat	12	Gg	N	43	0.09	2120		60	33	44	11	40	69	241
Tibbitt	13	G	N			115		19	10	7	3	1	49	67
Yellowknife	13	G	N	16300	1.2	71	-4	7	5	3	1	0.3	18	28
Wecho	13	G	N	3400	0.3	90	-4	9	6	6	2	1	20	37
Emile	13	G	N	4850	0.5	103	-4	12	9	4	2	0.2	36	49
Indin	13	G	N	1520	0.2	163	-4	8	5	3	2	0.2	16	33
Mc Crea	13	G	N			115	-4	8	5	5	2		21	33
Yellowknife	13	G	N	16300	1.2	71	-4	7	5	4	2	0.2	22	30
Cameroon	13	G	N	3630	0.2	49	-4	9	5	4	1	0.01	23	32
Tibbitt	13	G	N			115	-4	20	10	8	3		49	70
Gattineau	13	G	N	23100	11.8	512	4.5	33	13	13	6		84	110
Saguenay	13	G	N	73000	46.1	631	4.5	32	13	24	6		87	120

(Continued on next page)

Table B.4 – Continued

Locality	Reference	Lithology	Spot samples	Area km^2	Discharge km^3/yr	Runoff mm/yr	T °C	Ca*	Mg*	Na* kmol/km^2/yr	K*	Si	HCO$_3$	$\sum$ cations meq/m^2/yr
Peribonka	13	G	N	26900	19.3	717	4.5	20	10	15	6		57	80
Mistassini	13	G	N	9870	6.2	626	4.5	26	14	21	6		81	107
Saint Maurice	13	G	N	32100	17.0	531	4.5	16	9	30	5		55	85
Wessoneau	13	G	N			480	4.5	27	15	14	8	40	30	106
Saint Maurice	13	G	N	32100	17.0	531	4.5	23	12	11	11	40	19	91
La Croche	13	G	N	1850	1.1	612	4.5	32	13	18	16	48	31	124
Rivière des Bostonnais	13	G	N	9870	4.7	480	4.5	28	9	23	10	32	24	108
Mistassini	13	G	N	9320	5.8	626	4.5	27	15	11	16	43	31	112
Mistassibi	13	G	N			677	4.5	31	13	13	20	44	31	120
Rivière aux Rats	13	G	N			480	4.5	25	13	15	13	42	24	105
Solmyren	14	G	N	27	9.4×10^{-03}	348	0.3	30	14	11	3		10	100
Vuoddasbacken	14	G	N	42	1.6×10^{-02}	383	0.3	16	11	15	1			70
Lilla Tivsjön	14	G	N	12.8	2.6×10^{-03}	203		21	11	7	0.2			71
Loch Vale	15	G	N	8.6	5.2×10^{-03}	604		13	4	7	2	15		42
Barhalde	16	G	N			1350	3.9	33	10	67	9	118	21	160
Tsukuba 1986	17	G	N	0.67	4.9×10^{-04}	734	13.1	53	38	144	8	254		336
Tsukuba 1987	17	G	N	0.67	4.8×10^{-04}	711		41	35	126	2	254		280
Tsukuba 1989	17	G	N	0.67	7.2×10^{-04}	1075		37	54	219	12	375		411
Breidvikdalen	18	G	N	1.89	7.0×10^{-03}	3712	7.5	25	15	17	3			100
Panther Lake	19	G	N	1.24	8.9×10^{-04}	720	5	57	15	20	32			197
Lysina	20	G	N	0.27	1.1×10^{-04}	419	5	27	9	25	8	74		107
Schluchsee	21	G	N	0.11	2.2×10^{-04}	1974	5	39	10	91	23			212
Estibère	22	G	Y	7	7.9×10^{-03}	1125	5	103	1	27		76	240	234
Margeride	23	G	N	89	6.3×10^{-02}	705	11	39	39	70	13	123	188	239
Storbergsbacken	24	G	N	9.4	2.3×10^{-03}	247	-0.2	13	5	8	2	20		46
Guayana	25	G	Y			2338		11	3	35	37	245		99
Mengong	26	G	N	0.6	1.6×10^{-04}	263	24	6	6	13		37	9	6
Kama	26	G	N			350	24	3	7	7		51		26
Awout	27	G	Y			350	24	5	7	8		47		31
Fala	27	G	N			350	24	3	7	11	2	56		32
Nsono	27	G	N			350	24	6	8	11	3	51		43
Bissi	27	G	N			350	24	8	10	13	0.4	54		49

(Continued on next page)

Table B.4 – Continued

Locality	Reference	Lithology	Spot samples	Area km²	Discharge km²/yr	Runoff mm/yr	T °C	Ca*	Mg*	Na*	K*	Si	HCO₃	∑cations meq/m²/yr
										kmol/km²/yr				
Bives	27	G	N			350	24	3	5	7	1	47		24
Minsamba	27	G	N			350	24		5	3		42		12
Mengong	27	G	Y			350	24	1	4	6	3	41		20

References: 1-Gislason et al. [1996], 2-Gannoun et al. [2006], Georg et al. [2007], Vigier et al. [2006], 3-Yde et al. [2005], 4-Louvat and Allègre [1997], 5-Louvat and Allègre [1998], 6-Benedetti et al. [1994], 7-Dessert et al. [2001], 8-Gupta et al. [2011], 9-Jha et al. [2009], 10-Das et al. [2005], 11-Hosein [2002], Hosein et al. [2004], 12-Hagedorn and Hasholt [2004], 13-Millot et al. [2002], 14-Calles [1983], 15-Mast et al. [1990], 16-Stahr et al. [1980], 17-Hirata and Muraoka [1993], 18-Skartveit [1981], 19-April et al. [1986], 20-Kräm et al. [1997], 21-Feger et al. [1990], 22-Oliva et al. [2004], 23-Négrel [1999], 24-Land and Öhlander [2000], 25-Edmond et al. [1995], 26-Viers et al. [1997], 27-Viers et al. [2000]

Bibliography

S. M. Aciego, B. Bourdon, M. Lupker, and J. Rickli. A new procedure for separating and measuring radiogenic isotopes (U, Th, Pa, Ra, Sr, Nd, Hf) in ice cores. *Chem. Geol.*, 266: 194–204, 2009.

J. G. Acker and O. P. Bricker. The influence of pH on biotite dissolution and alteration kinetics at low temperature. *Geochim. Cosmochim. Acta*, 56:3073–3092, 1992.

M. Amini, A. Eisenhauer, F. Böhm, J. Fietzke, W. Bach, D. Garbe-Schönberg, M. Rosner, B. Bock, K. S. Lackschewitz, and F. Hauff. Calcium isotope ($\delta^{44/40}$Ca) fractionation along hydrothermal pathways, Logatchev field (Mid-Atlantic Ridge, 14°45'N). *Geochim. Cosmochim. Acta*, 72: 4107–4122, 2008.

M. Amini, A. Eisenhauer, F. Böhm, C. Holmden, K. Kreissig, F. Hauff, and K. P. Jochum. Calcium isotopes ($\delta^{40/44}$Ca) in MPI-DING reference glasses, USGS rock powders and various rocks: Evidence for Ca isotope fractionation in terrestrial silicates. *Geostand. Geoanal. Res.*, 33:231–247, 2009.

R. Amundson. Soil formation. In H. D. Holland and K. K. Turekian, editors, *Treatise on Geochemistry*, volume 5: Surface and ground water, weathering and soils. Elsevier, 2003.

S. P. Anderson, J. I. Drever, and N. F. Humphrey. Chemical weathering in glacial environments. *Geology*, 25:399–402, 1997a.

S. P. Anderson, W. E. Dietrich, D. R. Montgomery, R. Torres, M. E. Conrad, and K. Loague. Subsurface flow paths in a steep, unchanneled catchment. *Water Resour. Res.*, 33:2637–2653, 1997b.

S. P. Anderson, J. I. Drever, C. D. Frost, and P. Holden. Chemical weathering in the foreland of a retreating glacier. *Geochim. Cosmochim. Acta*, 64:1173–1189, 2000.

R. April, R. Newton, and L. Truettner Coles. Chemical weathering in two Adirondack watersheds: past and present-day rates. *Geol. Soc. Am. Bull.*, 97:1232–1238, 1986.

K. Arn, R. Hosein, K. B. Föllmi, P. Steinmann, A. Aubert, and J. Kramers. Strontium isotope systematics in two glaciated crystalline catchments: Rhone and Oberaar glaciers (Swiss Alps). *Schweiz. Mineral. Petrogr. Mitt.*, 83:273–283, 2003.

D. Aubert, A. Probst, P. Stille, and D. Viville. Evidence of hydrological control of Sr behavior in stream water (Strengbach catchment, Vosges mountains, France). *Appl. Geochem.*, 17:285–300, 2002.

S. W. Bailey, J. W. Hornbeck, C. T. Driscoll, and H. E. Gaudette. Calcium inputs and transport in a base-poor forest ecosystem as interpreted by Sr isotopes. *Water Resour. Res.*, 32:707–719, 1996.

D. C. Bain, A. Mellor, M. S. E. Robertson-Rintoul, and S. T. Buckland. Variations in weathering processes and rates with time in a chronosequence of soils from Glen Feshie, Scotland. *Geoderma*, 57:275–293, 1993.

L. S. Balistrieri, D. M. Borrok, R. B. Wanty, and W. I. Ridley. Fractionation of Cu and Zn isotopes during adsorption onto amorphous Fe(III) oxyhydroxide: Experimental mixing of acid rock drainage and ambient river water. *Geochim. Cosmochim. Acta*, 72:311–328, 2008.

Z. Balogh-Brunstad, C. K. Keller, B. T. Bormann, R. O'Brien, D. Wang, and G. Hawley. Chemical weathering and chemical denudation dynamics through ecosystem development and disturbance. *Global Biogeochem. Cycles*, 22:GB1007, 2008.

I. Bartholomew, P. Nienow, A. Sole, D. Mair, T. Cowton, S. Palmer, and J. Wadham. Supraglacial forcing of subglacial drainage in the ablation zone of the Greenland ice sheet. *Geophys. Res. Lett.*, 38:L08502, 2011.

N. Bélanger and C. Holmden. Influence of landscape on the apportionment of Ca nutrition in a Boreal Shield forest of Saskatchewan (Canada) using $^{87}Sr/^{86}Sr$ as a tracer. *Can. J. Soil Sci.*, 90:267–288, 2010.

M. F. Benedetti, O. Menard, Y. Noack, A. Carvalho, and D. Nahon. Water-rock interactions in tropical catchments: field rates of weathering and biomass impact. *Chem. Geol.*, 118:203–220, 1994.

M. R. Bennett and N. F. Glasser. *Glacial geology - Ice sheets and landforms*. Wiley, 1996.

S. M. Bernasconi, I. Christl, I. Hajdas, S. Zimmermann, F. Hagedorn, R. H. Smittenberg, G. Furrer, J. Zeyer, I. Brunner, B. Frey, M. Plötze, A. Lapanje, P. Edwards, H. Olde Venterink, H. Göransson, E. Frossard, E. Bünemann, J. Jansa, F. Tamburini, M. Welc, E. Mitchell, B. Bourdon, R. Kretzschmar, B. Reynolds, E. Lemarchand, J. Wiederhold, E. Tipper, M. Kiczka, R. Hindshaw, M. Stähli, T. Jonas, J. Magnusson, A. Bauder, D. Farinotti, M. Huss, L. Wacker, and K. Abbaspour. Weathering, soil formation and initial ecosystem evolution on a glacier forefield: a case study from the Damma Glacier, Switzerland. *Miner. Mag.*, 72:19–22, 2008.

R. A. Berner and Z. Kothavala. GEOCARB III: A revised model of atmospheric CO_2 over phanerozoic time. *Am. J. Sci.*, 301:182–204, 2001.

R. A. Berner, A. C. Lasaga, and R. M. Garrels. The carbonate-silicate geochemical cycle and its effect on atmospheric carbon dioxide over the past 100 million years. *Am. J. Sci.*, 283:641–683, 1983.

R. A. Berner, J.-L. Rao, S. Chang, R. O'Brien, and C. K. Keller. Seasonal variability of adsorption and exchange equilibria in soil waters. *Aquat. Geochem.*, 4:273–290, 1998.

J. L. Birck and C. J. Allègre. Chronology and chemical history of the parent body of basaltic achondrites by the ^{87}Rb-^{87}Sr method. *Earth Planet. Sci. Lett.*, 39:37–51, 1978.

J. R. Black, Q-Z. Yin, J. R. Rustad, and W. H. Casey. Magnesium isotopic equilibrium in chlorophylls. *J. Am. Chem. Soc.*, 129:8690–8691, 2007.

J. R. Black, E. Epstein, W. D. Rains, Q-Z. Yin, and W. H. Casey. Magnesium-isotope fractionation during plant growth. *Environ. Sci. Technol.*, 42:7831–7836, 2008.

J. D. Blum and Y. Erel. Rb-Sr isotope systematics of a granitic soil chronosequence: The importance of biotite weathering. *Geochim. Cosmochim. Acta*, 61:3193–3204, 1997.

J. D. Blum, Y. Erel, and K. Brown. $^{87}Sr/^{86}Sr$ ratios of Sierra Nevada stream waters: Implications for relative mineral weathering rates. *Geochim. Cosmochim. Acta*, 58:5019–5025, 1994.

J. D. Blum, A. Klaue, C. A. Nezat, C. T. Driscoll, C. E. Johnson, T. G. Siccama, C. Eagar, T. J. Fahey, and G. E. Likens. Mycorrhizal weathering of apatite as an important calcium source in base-poor forest ecosystems. *Nature*, 417:729–731, 2002.

J. D. Blum, A. A. Dasch, S. P. Hamburg, R. D. Yanai, and M. A. Arthur. Use of foliar Ca/Sr discrimination and $^{87}Sr/^{86}Sr$ ratios to determine soil Ca sources to sugar maple foliage in a northern hardwood forest. *Biogeochemistry*, 87:287–296, 2008.

G. J. S. Bluth and L. R. Kump. Lithologic and climatologic controls of river chemistry. *Geochim. Cosmochim. Acta*, 58:2341–2359, 1994.

F. Böhm, N. Gussone, A. Eisenhauer, W-C. Dullo, S. Reynaud, and A. Paytan. Calcium isotope fractionation in modern scleractinian corals. *Geochim. Cosmochim. Acta*, 70:4452–4462, 2006.

E. B. Bolou-Bi, A. Poszwa, C. Leyval, and N. Vigier. Experimental determination of magnesium isotope fractionation during higher plant growth. *Geochim. Cosmochim. Acta*, 74:2523–2537, 2010.

B. T. Bormann, D. Wang, F. H. Bormann, G. Benoit, R. April, and M. C. Snyder. Rapid, plant-induced weathering in an aggrading experimental ecosystem. *Biogeochemistry*, 43:129–155, 1998.

D. J. Bottomley, D. Craig, and L. M. Johnston. Oxygen-18 studies of snowmelt runoff in a small precambrian shield watershed: Implications for streamwater acidification in acid-sensitive terrain. *J. Hydrol.*, 88:213–234, 1986.

S. F. Boulyga. Calcium isotope analysis by mass spectrometry. *Mass Spec. Rev.*, 29:685–716, 2010.

I. C. Bourg, F. M. Richter, J. N. Christensen, and G. Sposito. Isotopic mass dependence of metal cation diffusion coefficients in liquid water. *Geochim. Cosmochim. Acta*, 74:2249–2256, 2010.

P. V. Brady and J. V. Walther. Controls on silicate dissolution rates in neutral and basic pH solutions at 25°C. *Geochim. Cosmochim. Acta*, 53:2823–2830, 1989.

S. L. Brantley. Reaction kinetics of primary rock-forming minerals under ambient conditions. In H. D. Holland and K. K. Turekian, editors, *Treatise on Geochemistry*, volume 5: Surface and ground water, weathering and soils. Elsevier, 2003.

S. L. Brantley, J. T. Chesley, and L. L. Stillings. Isotopic ratios and release rates of strontium measured from weathering feldspars. *Geochim. Cosmochim. Acta*, 62:1493–1500, 1998.

S. L. Brantley, T. S. White, A. F. White, D. Sparks, D. Richter, K. Pregitzer, L. Derry, J. Chorover, O. Chadwick, R. April, S. Anderson, and R. Amundson. Frontiers in exploration of the Critical Zone: Report of a workshop sponsored by the National Science Foundation (NSF), October 24-26 2005.

E. J. Brook, T. Sowers, and J. Orchardo. Rapid variations in atmospheric methane concentration during the past 110,000 years. *Science*, 273:1087–1091, 1996.

G. H. Brown. Glacier meltwater hydrochemistry. *Appl. Geochem.*, 17:855–883, 2002.

L. E. Brown, D. M. Hannah, A. M. Milner, C. Soulsby, A. J. Hodson, and M. J. Brewer. Water source dynamics in a glacierized alpine river basin (Taillon-Gabiétous, French Pyrénées). *Water Resour. Res.*, 42:W08404, 2006.

T. Bullen, A. White, A. Blum, J. Harden, and M. Schulz. Chemical weathering of a soil chronosequence on granitoid alluvium: II. mineralogic and isotopic constraints on the behavior of strontium. *Geochim. Cosmochim. Acta*, 61:291–306, 1997.

T. D. Bullen and S. W. Bailey. Identifying calcium sources at an acid deposition-impacted spruce forest: a strontium isotope, alkaline earth element multi-tracer approach. *Biogeochemistry*, 74: 63–99, 2005.

T. D. Bullen, D. P. Krabbenhoft, and C. Kendall. Kinetic and mineralogic controls on the evolution of groundwater chemistry and $^{87}Sr/^{86}Sr$ in a sandy silicate aquifer, northern Wisconsin, USA. *Geochim. Cosmochim. Acta*, 60:1807–1821, 1996.

T. D. Bullen, J. A. Fitzpatrick, A. F. White, M. S. Schulz, and D. V. Vivit. Calcium stable isotope evidence for three soil calcium pools at a granitoid chronosequence. In R. B. Wanty and R. R. Seal II, editors, *Water-Rock Interaction, Proceedings of the Eleventh International Symposium on Water-Rock Interaction, Saratoga Springs, New York, July 2004*, volume 1, pages 813–817. Taylor & Francis, London, 2004.

J. M. Buttle. Isotope hydrograph separations and rapid delivery of pre-event water from drainage basins. *Prog. Phys. Geog.*, 18:16–41, 1994.

U. M. Calles. Dissolved inorganic substances: A study of mass balance in three small drainage basins. *Hydrobiologia*, 101:13–18, 1983.

L. Camarero and J. Catalan. Variability in the chemistry of precipitation in the Pyrenees (northeastern Spain): Dominance of storm origin and lack of altitude influence. *J. Geophys. Res.*, 101:29491–29498, 1996.

E. M. Cameron, G. E. M. Hall, J. Veizer, and H. R. Krouse. Isotopic and elemental hydrogeochemistry of a major river system: Fraser River, British Columbia, Canada. *Chem. Geol.*, 122:149–169, 1995.

R. C. Capo, B. W. Stewart, and O. A. Chadwick. Strontium isotopes as tracers of ecosystem processes: theory and methods. *Geoderma*, 82:197–225, 1998.

G. Caro, D. A. Papanastassiou, and G. J. Wasserburg. ^{40}K-^{40}Ca isotopic constraints on the oceanic calcium cycle. *Earth Planet. Sci. Lett.*, 296:124–132, 2010.

O. Carugo, K. Djinović, and M. Rizzi. Comparison of the co-ordinative behaviour of calcium(II) and magnesium(II) from crystallographic data. *J. Chem. Soc. Dalton Trans.*, 14:2127–2135, 1993.

B. Cenki-Tok, F. Chabaux, D. Lemarchand, A-D. Schmitt, M-C. Pierret, D. Viville, and P. Stille. The impact of water-rock interaction and vegetation on calcium isotope fractionation in soil- and stream waters of a small, forested catchment (the Strengbach case). *Geochim. Cosmochim. Acta*, 73:2215–2228, 2009.

V. T.-C. Chang, R. J. P. Williams, A. Makishima, N. S. Belshaw, and R. K. O'Nions. Mg and Ca isotope fractionation during $CaCO_3$ biomineralisation. *Biochem. Biophy. Res. Commun.*, 323: 79–85, 2004.

N.-C. Chu, G. M. Henderson, N. S. Belshaw, and R. E. M. Hedges. Establishing the potential of Ca isotopes as proxy for consumption of dairy products. *Appl. Geochem.*, 21:1656–1667, 2006.

D. W. Clow and J. I. Drever. Weathering rates as a function of flow through an alpine soil. *Chem. Geol.*, 132:131–141, 1996.

D. W. Clow and M. A. Mast. Mechanisms for chemostatic behavior in catchments: Implications for CO_2 consumption by mineral weathering. *Chem. Geol.*, 269:40–51, 2010.

D. W. Clow, M. A. Mast, T. D. Bullen, and J. T. Turk. Strontium 87/strontium 86 as a tracer of mineral weathering reactions and calcium sources in an alpine/subalpine watershed, Loch Vale, Colorado. *Water Resour. Res.*, 33:1335–1351, 1997.

D. N. Collins. Solute yield from a glacierized high mountain basin. In *Dissolved loads of rivers and surface water quantity/quality relationships. Proceeding of the Humburg Symposium, August 1983*, IAHS Publ. no. 141, 1983.

J. T. Corless. Observations on the isotopic geochemitry of calcium. *Earth Planet. Sci. Lett.*, 4: 475–478, 1968.

A. Das, S. Krishnaswami, M. M. Sarin, and K. Pande. Chemical weathering in the Krishna Basin and Weatern Ghats of the Deccan Traps, India: Rates of basalt weathering and their controls. *Geochim. Cosmochim. Acta*, 69:2067–2084, 2005.

A. A. Dasch, J. D. Blum, C. Eagar, T. J. Fahey, C. T. Driscoll, and T. G. Siccama. The relative uptake of Ca and Sr into tree foliage using a whole-watershed calcium addition. *Biogeochemistry*, 80:21–41, 2006.

M. De Angelis and A. Gaudichet. Saharan dust deposition over Mont Blanc (French Alps) during the last 30 years. *Tellus*, 43B:61–75, 1991.

C. L. De La Rocha and D. J. DePaolo. Isotopic evidence for variations in the marine calcium cycle over the Cenozoic. *Science*, 289:1176–1178, 2000.

G. de Souza. $^{87}Sr/^{86}Sr$ and $\delta^{88/86}Sr$ in the forefield of the Damma Glacier, Switzerland. Master's thesis, ETH Zurich, 2007.

G. F. de Souza, B. C. Reynolds, M. Kiczka, and B. Bourdon. Evidence for mass-dependent isotopic fractionation of strontium in a glaciated granitic watershed. *Geochim. Cosmochim. Acta*, 74: 2596–2614, 2010.

T. J. Dempster. Isotope systematics in minerals: biotite rejuvenation and exchange during Alpine metamorphism. *Earth Planet. Sci. Lett.*, 78:355–367, 1986.

C. Deniel and C. Pin. Single-stage method for the simultaneous isolation of lead and strontium from silicate samples for isotopic measurements. *Anal. Chim. Acta*, 426:95–103, 2001.

D. J. DePaolo. Calcium isotopic variations produced by biological, kinetic, radiogenic and nucleosynthetic processes. In C. M. Johnson, B. L. Beard, and F. Albarède, editors, *Geochemistry of Non-traditional Stable Isotopes, Reviews in Mineralogy and Geochemistry*, volume 55, pages 255–288. Mineralogical Society of America, Washington D.C., 2004.

C. Dessert, B. Dupré, L. M. François, J. Schott, J. Gaillardet, G. Chakrapani, and S. Bajpai. Erosion of Deccan Traps determined by river geochemistry: impact on the global climate and the $^{87}Sr/^{86}Sr$ ratio of seawater. *Earth Planet. Sci. Lett.*, 188:459–474, 2001.

C. Dessert, B. Dupré, J. Gaillardet, L. M. François, and C. J. Allègre. Basalt weathering laws and the impact of basalt weathering on the global carbon cycle. *Chem. Geol.*, 202:257–273, 2003.

A. R. Doaigey. Occurence, type, and location of calcium oxalate crstals in leaves and stems of 16 species of poisonous plants. *Am. J. Bot.*, 78:1608–1616, 1991.

J. I. Drever. The effect of land plants on weathering rates of silicate minerals. *Geochim. Cosmochim. Acta*, 58:2325–2332, 1994.

J. I. Drever. *The Geochemistry of Natural Waters*. Prentice-Hall, Upper Saddle River, N.J., 3rd edition, 1997.

M. C. Drew and O. Biddulph. Effect of metabolic inhibitors and temperature on uptake and translocation of ^{45}Ca and ^{42}K by intact bean plants. *Plant Physiol.*, 48:426–432, 1971.

T. Drouet and J. Herbauts. Evaluation of the mobility and discrimination of Ca, Sr and Ba in forest ecosystems: consequence on the use of alkaline-earth element ratios as tracers of Ca. *Plant Soil*, 302:105–124, 2008.

J. M. Edmond and Y. Huh. Chemical weathering yields from basement and orogenic terrains in hot and cold climates. In W. F. Ruddiman, editor, *Tectonic uplift and climate change*, pages 329–351, New York, 1997. Plenum Press.

J. M. Edmond, M. R. Palmer, C. I. Measures, B. Grant, and R. F. Stallard. The fluvial geochemistry and denudation rate of the Guayana Shield in Venezuela, Colombia, and Brazil. *Geochim. Cosmochim. Acta*, 59:3301–3325, 1995.

M. Egli, P. Fitze, and A. Mirabella. Weathering and evolution of soils formed on granitic, glacial deposits: results from chronosequences of Swiss alpine environments. *Catena*, 45:19–47, 2001.

H. Einspahr and C. E. Bugg. The geometry of calcium-carboxylate interactions in crystalline complexes. *Acta Cryst.*, B37:1044–1052, 1981.

A. Eisenhauer, T. F. Nägler, P. Stille, J. Kramers, N. Gussone, B. Bock, J. Fietzke, D. Hippler, and A-D. Schmitt. Proposal for international agreement on Ca notation resulting from discussions at workshops on stable isotope meaurements held in Davos (Goldschmidt 2002) and Nice (EGS-AGU-EUG 2003). *Geostand. Geoanal. Res.*, 28:149–151, 2004.

E. Engström, I. Rodushkin, J. Ingri, D. C. Baxter, F. Ecke, H. Österlund, and B. Öhlander. Temporal isotopic variations of dissolved silicon in a pristine boreal river. *Chem. Geol.*, 271: 142–152, 2010.

E. Epstein and J. E. Leggett. The absorption of alkaline earth cations by barley roots: kinetics and mechanism. *Am. Jour. Bot.*, 41:785–791, 1954.

Y. Erel, J. D. Blum, E. Roueff, and J. Ganor. Lead and strontium isotopes as monitors of experimental granitoid mineral dissolution. *Geochim. Cosmochim. Acta*, 68:4649–4663, 2004.

D. E. Evans, S. A. Briars, and L. E. Williams. Active calcium transport by plant cell membranes. *J. Exp. Bot.*, 42:285–303, 1991.

S. A. Ewing, W. Yang, D. J. DePaolo, G. Michalski, C. Kendall, B. W. Stewart, M. Thiemens, and R. Amundson. Non-biological fractionation of stable Ca isotopes in soils of the Atacama Desert, Chile. *Geochim. Cosmochim. Acta*, 72:1096–1110, 2008.

I. J. Fairchild, J. A. Killawee, M. J. Sharp, B. Spiro, B. Hubbard, R. D. Lorrain, and J.-L. Tison. Solute generation and transfer from a chemically reactive alpine glacial-proglacial system. *Earth Surf. Process. Landforms*, 24:1189–1211, 1999.

M. S. Fantle and T. D. Bullen. Essentials of iron, chromium and calcium isotope analysis of natural materials by thermal ionization mass spectrometry. *Chem. Geol.*, 258:50–64, 2009.

M. S. Fantle and D. J. DePaolo. Variations in the marine Ca cycle over the past 20 million years. *Earth Planet. Sci. Lett.*, 237:102–117, 2005.

M. S. Fantle and D. J. DePaolo. Ca isotopes in carbonate sediment and pore fluid from ODP Site 807a: The Ca^{2+}(aq)-calcite equilibrium fractionation factor and calcite recrystallization rates in Pleistocene sediments. *Geochim. Cosmochim. Acta*, 71:2524–2546, 2007.

D. Farinotti, J. Magnusson, M. Huss, and A. Bauder. Snow accumulation distribution inferred from time-lapse photography and simple modelling. *Hydrol. Process.*, 24:2087–2097, 2010.

J. Farkaš, F. Böhm, K. Wallmann, J. Blenkinsop, A. Eisenhauer, R. van Geldern, A. Munnecke, S. Voigt, and J. Veizer. Calcium isotope record of Phanerozoic oceans: Implications for chemical evolution of seawater and its causative mechanisms. *Geochim. Cosmochim. Acta*, 71:5117–5134, 2007.

K. H. Feger, G. Brahmer, and H. W. Zöttl. Element budgets of two contrasting catchments in the Black Forest (Federal Republic of Germany). *J. Hydrol.*, 116:85–99, 1990.

I. B. Ferguson and E. G. Bollard. The movement of calcium in woody stems. *Ann. Bot.*, 40:1057–1065, 1976.

J. Fietzke, A. Eisenhauer, N. Gussone, B. Bock, V. Liebetrau, T. F. Nägler, H. J. Spero, J. Bijma, and C. Dullo. Direct measurement of $^{44}Ca/^{40}Ca$ ratios by MC-ICP-MS using the cool plasma technique. *Chem. Geol.*, 206:11–20, 2004.

J. B. Finley and J. I. Drever. Chemical mass balance and rates of mineral weathering in a high-elevation catchment, West Glacier Lake, Wyoming. *Hydrol. Process.*, 11:745–764, 1997.

I. R. Fletcher, N. J. McNaughton, R. T. Pidgeon, and K. J. R. Rosman. Sequential closure of K-Ca and Rb-Sr isotopic systems in Archaean micas. *Chem. Geol.*, 138:289–301, 1997.

A. G. Fountain. Effect of snow and firn hydrology on the physical and chemical characteristics of glaical runoff. *Hydrol. Process.*, 10:509–521, 1996.

C. France-Lanord, M. Evans, J-E. Hurtrez, and J. Riotte. Annual dissolved fluxes from Central Nepal rivers: budget of chemical erosion in the Himalayas. *C. R. Geosci.*, 335:1131–1140, 2003.

R. Freydier, B. Dupré, J-L. Dandurand, J-P. Fortune, and L. Sigha-Nkamdjou. Trace elements and major species in precipitation at African stations: concentrations and sources. *Bull. Soc. géol. France*, 173:129–146, 2002.

E. J. Gabet, D. Wolff-Boenisch, H. Langner, D. Burbank, and J. Putkonen. Geomorphic and climatic controls on chemical weathering in the High Himalayas of Nepal. *Geomorphology*, 122:205–210, 2010.

J. Gaillardet, B. Dupré, P. Louvat, and C. J. Allègre. Gobal silicate weathering and CO_2 consumption rates deduced from the chemistry of large rivers. *Chem. Geol.*, 159:3–30, 1999.

A. Galy and C. France-Lanord. Weathering processes in the Ganges-Brahmaputra basin and the riverine alkalinity budget. *Chem. Geol.*, 159:31–60, 1999.

A. Gannoun, K. W. Burton, N. Vigier, S. Gíslason, N. Rogers, F. Mokadem, and B. Sigfússon. The influence of weathering process on riverine osmium isotopes in a basaltic terrain. *Earth Planet. Sci. Lett.*, 243:732–748, 2006.

R. M. Garrels and F. T. Mackenzie. Origin of the chemical compositions of some springs and lakes. In R. F. Gould, editor, *Equilibrium concepts in natural water systems*, volume 67 of *Advances in Chemistry Series*, pages 222–242. American Chemical Society, Washington DC, 1967.

J. R. Gat. Oxygen and hydrogen isotopes in the hydrologic cycle. *Annu. Rev. Earth Planet. Sci.*, 24:225–262, 1996.

R. B. Georg, B. C. Reynolds, A. J. West, K. W. Burton, and A. N. Halliday. Silicon isotope variations accompanying basalt weathering in Iceland. *Earth Planet. Sci. Lett.*, 261:476–490, 2007.

M. T. Gibbs and L. R. Kump. Global chemical erosion during the last glacial maximum and the present: Sensitivity to changes in lithology and hydrology. *Paleooceanography*, 9:529–543, 1994.

M. Giovannetti and B. Mosse. An evaluation of techniques for measuring vesicular arbuscular mycorrhizal infection in roots. *New Phytol.*, 84:489–500, 1980.

S. R. Gíslason, S. Arnórsson, and H. Ármannsson. Chemical weathering of basalt in southwest Iceland: Effects of runoff, age of rocks and vegetative/glacial cover. *Am. J. Sci.*, 296:837–907, 1996.

S. R. Gíslason, E. H. Oelkers, E. S. Eiriksdottir, M. I. Kardjilov, G. Gisladottir, B. Sigfusson, A. Snorrason, S. Elefsen, J. Hardardottir, P. Torssander, and N. Oskarsson. Direct evidence of the feedback between climate and weathering. *Earth Planet. Sci. Lett.*, 277:213–222, 2009.

S. E. Godsey, J. W. Kirchner, and D. W. Clow. Concentration-discharge relationships reflect chemostatic characteristics of US catchments. *Hydrol. Process.*, 23:1844–1864, 2009.

K. Gopalan, D. Macdougall, and C. Macisaac. Evaluation of a ^{42}Ca-^{43}Ca double spike for high precision Ca isotope analysis. *Int. J. Mass Spectrom.*, 248:9–16, 2006.

K. Gopalan, J. D. Macdougall, and C. Macisaac. High precision determination of $^{48}Ca/^{42}Ca$ ratio by TIMS for Ca isotope fractionation studies. *Geostand. Geoanal. Res.*, 31:227–236, 2007.

E. Gorham, P. M. Vitousek, and W. A. Reiners. The regulation of chemical budgets over the course of terrestrial ecosystem succession. *Ann. Rev. Ecol. Syst.*, 10:53–84, 1979.

A. S. Goudie and N. J. Middleton. Saharan dust storms: nature and consequences. *Earth Sci. Rev.*, 56:179–204, 2001.

M. M. Guha and R. L. Mitchell. The trace and major element composition of the leaves of some deciduous trees II. Seasonal changes. *Plant and Soil*, 24:90–112, 1966.

H. Gupta, G. J. Chakrapani, K. Selvaraj, and S.-J. Kao. The fluvial geochemistry, contributions of silicate, carbonate and saline-alkaline components to chemical weathering flux and controlling parameters: Narmada River (Deccan Traps), India. *Geochim. Cosmochim. Acta*, 75:800–824, 2011.

N. Gussone, A. Eisenhauer, A. Heuser, M. Dietzel, B. Bock, F. Böhm, H. J. Spero, D. W. Lea, J. Bijma, and T. F. Nägler. Model for kinetic effects on calcium isotope fractionation ($\delta\,^{44}Ca$) in inorganic aragonite and cultured planktonic foraminifera. *Geochim. Cosmochim. Acta*, 67: 1375–1382, 2003.

N. Gussone, G. Langer, S. Thoms, G. Nehrke, A. Eisenhauer, U. Riebesell, and G. Wefer. Cellular calcium pathways and isotope fractionation in *Emiliania huxleyi*. *Geology*, 34:625–628, 2006.

B. Hagedorn and B. Hasholt. Hydrology, geochemistry and Sr isotopes in solids and solutes of the meltwater from Mittivakkat Gletscher, SE Greenland. *Nord. Hydrol.*, 35:369–380, 2004.

B. Hallet, L. Hunter, and J. Bogen. Rates of erosion and sediment evacuation by glaciers: A review of field data and their implications. *Global Planet. Change*, 12:213–235, 1996.

T. M. Harrison, M. T. Heizler, K. D. McKeegan, and A. K. Schmitt. In situ ^{40}K-^{40}Ca 'double-plus' SIMS dating resolves Klokken feldspar ^{40}K-^{40}Ar paradox. *Earth Planet. Sci. Lett.*, 299: 426–433, 2010.

R. J. Haynes. Ion exchange properties of roots and ionic interactions within the root apoplasm: their role in ion accumulation by plants. *Bot. Rev.*, 46:75–99, 1980.

A. Heuser and A. Eisenhauer. The calcium isotope composition ($\delta^{44/40}$Ca) of NIST SRM 915b and NIST SRM 1486. *Geostand. Geoanal. Res.*, 32:311–315, 2008.

A. Heuser and A. Eisenhauer. A pilot study on the use of natural calcium isotope (^{44}Ca/^{40}Ca) fractionation in urine as a proxy for the human body calcium balance. *Bone*, 46:889–896, 2010.

A. Heuser, T. Tütken, N. Gussone, and S. J. G. Galer. Calcium isotopes in fossil bones and teeth - diagenetic versus biogenic origin. *Geochim. Cosmochim. Acta*, 75:3419–3433, 2011.

R. S. Hindshaw, B. C. Reynolds, J. G. Wiederhold, R. Kretzschmar, and B. Bourdon. Calcium isotopes in a proglacial weathering environment: Damma glacier, Switzerland. *Geochim. Cosmochim. Acta*, 75:106–118, 2011a.

R. S. Hindshaw, E. T. Tipper, B. C. Reynolds, E. Lemarchand, J. G. Wiederhold, J. Magnusson, S. M. Bernasconi, R. Kretzschmar, and B. Bourdon. Hydrological control of stream water chemistry in a glacial catchment (Damma Glacier, Switzerland). *Chem. Geol.*, 285:215–230, 2011b.

D. Hippler, A-D. Schmitt, N. Gussone, A. Heuser, P. Stille, A. Eisenhauer, and T. F. Nägler. Calcium isotopic composition of various reference materials and seawater. *Geostand. Geoanal. Res.*, 27:13–19, 2003.

T. Hirata and K. Muraoka. The relation between water migration and chemical processes in a forest ecosystem. In N. E. Peters, E. Hoehn, C. Leibundgut, N. Tase, and D. E. Walling, editors, *Tracers in Hydrology*, number 215, pages 31–40, Wallingford, 1993. IAHS Press.

A. R. Hoch, M. M. Reddy, and J. I. Drever. Importance of mechanical disaggregation in chemical weathering in a cold alpine environment, San Juan Mountains, Colorado. *Geology*, 111:304–314, 1999.

R. Hodgkins, M. Tranter, and J. A. Dowdeswell. Solute provenance, transport and denudation in a high arctic glacierized catchment. *Hydrol. Process.*, 11:1813–1832, 1997.

R. Hodgkins, M. Tranter, and J. A. Dowdeswell. The hydrochemistry of runoff from a 'cold-based' glacier in the High Arctic (Scott Turnerbreen, Svalbard). *Hydrol. Process.*, 12:87–103, 1998.

R. Hodgkins, R. Cooper, J. Wadham, and M. Tranter. The hydrology of the proglacial zone of a high-Arctic glacier (Finsterwalderbreen, Svalbard): Atmospheric and surface water fluxes. *J. Hydrol.*, 378:150–160, 2009.

A. Hodson, M. Tranter, and G. Vatne. Contemporary rates of chemical denudation and atmospheric CO_2 sequestration in glacier basins: An Arctic perspective. *Earth Surf. Process. Landforms*, 25:1447–1471, 2000.

J. F. Hogan and J. D. Blum. Tracing hydrologic flow paths in a small forested watershed using variations in ^{87}Sr/^{86}Sr, [Ca]/[Sr], [Ba]/[Sr] and δ^{18}O. *Water Resour. Res.*, 39:1282, 2003.

H. D. Holland. *The chemistry of the atmosphere and oceans*. Wiley-Interscience, New York, 1978.

C. Holmden. Measurement of δ^{44}Ca using a ^{43}Ca-^{42}Ca double-spike TIMS technique. In *Summary of Investigations*, volume 1. Saskatchewan Geological Survey, 2005.

C. Holmden. Ca isotope study of Ordovician dolomite, limestone and anhydrite in the Williston Basin: Implications for subsurface dolomitization and local Ca cycling. *Chem. Geol.*, 268: 180–188, 2009.

C. Holmden and N. Bélanger. Ca isotope cycling in a forested ecosystem. *Geochim. Cosmochim. Acta*, 74:995–1015, 2010.

E. Hose, D. T. Clarkson, E. Steudle, L. Schreiber, and W. Hartung. The exodermis: a variable apoplastic barrier. *J. Exp. Bot.*, 52:2245–2264, 2001.

R. Hosein. *Biogeochemical weathering processes in the glacierised Rhône and Oberaar catchments, Switzerland, and the Apure catchment, Venezuela*. PhD thesis, Université de Neuchâtel, 2002.

R. Hosein, K. Arn, P. Steinmann, T. Adatte, and K. B. Föllmi. Carbonate and silicate weathering in two presently glaciated, crystalline catchments in the Swiss Alps. *Geochim. Cosmochim. Acta*, 68:1021–1033, 2004.

M. T. Hren, C. P. Chamberlain, G. E. Hilley, P. M. Blisniuk, and B. Bookhagen. Major ion chemistry of the Yarlung Tsangpo-Brahmaputra river: Chemical weathering, erosion, and CO_2 consumption in the southern Tibetan plateau and eastern syntaxis of the Himalaya. *Geochim. Cosmochim. Acta*, 71:2907–2935, 2007.

S. Huang, J. Farkaš, and S. B. Jacobsen. Calcium isotopic fractionation between clinopyroxene and orthopyroxene from mantle peridotites. *Earth Planet. Sci. Lett.*, 292:337–344, 2010.

R. J. Huggett. Soil chronosequences, soil development, and soil evolution: a critical review. *Catena*, 32:155–172, 1998.

Y. Huh, G. Panteleyev, D. Babich, A. Zaitsev, and J. M. Edmond. The fluvial geochemistry of the rivers of Eastern Siberia: II. Tributaries of the Lena, Omoloy, Yana, Indigirka, Kolyma, and Anadyr draining the collisional/accretionary zone of the Verkhoyansk and Cherskiy ranges. *Geochim. Cosmochim. Acta*, 62:2053–2075, 1998.

P. Huybrechts. Basal temperature conditions of the Greenland ice sheet during the glacial cycles. *Ann. Glaciol.*, 23:226–236, 1996.

J. Imbrie, E. A. Boyle, S. C. Clemens, A. Duffy, W. R. Howard, G. Kukla, J. Kutzbach, D. G. Martinson, A. McIntyre, A. C. Mix, B. Molfino, J. J. Morley, L. C. Peterson, N. G. Pisias, W. L. Prell, M. E. Raymo, N. J. Shackleton, and J. R. Toggweiler. On the structure and origin of major glaciation cycles 1. Linear responses to Milankovitch forcing. *Paleoceanography*, 7:701–738, 1992.

J. Imbrie, A. Berger, E. A. Boyle, S. C. Clemens, A. Duffy, W. R. Howard, G. Kukla, J. Kutzbach, D. G. Martinson, A. McIntyre, A. C. Mix, B. Molfino, J. J. Morley, L. C. Peterson, N. G. Pisias, W. L. Prell, M. E. Raymo, N. J. Shackleton, and J. R. Toggweiler. On the structure and origin of major glaciation cycles 2. The 100,000 year cycle. *Paleoceanography*, 8:699–735, 1993.

A. D. Jacobsen, J. D. Blum, C. P. Chamberlain, M. A. Poage, and V. F. Sloan. Ca/Sr and Sr isotope systematics of a Himalayan glacial chronosequence: Carbonate versus silicate weathering rates as a function of landscape surface age. *Geochim. Cosmochim. Acta*, 66:13–27, 2002.

A. D. Jacobson and C. Holmden. δ^{44}Ca evolution in a carbonate aquifer and its bearing on the equilibrium isotope fractionation factor for calcite. *Earth Planet. Sci. Lett.*, 270:349–353, 2008.

F. Jalilehvand, D. Spångberg, P. Lindqvist-Reis, K. Hermansson, I. Persson, and M. Sandström. Hydration of the calcium ion. An EXAFS, large-angle X-ray scattering, and molecular dynamics simulation study. *J. Am. Chem. Soc.*, 123:431–441, 2001.

H. Jenny. *Factors of soil formation: a system of quantitative pedology*. McGraw-Hill, New York, 1941.

P. K. Jha, J. Tiwari, U. K. Singh, M. Kumar, and V. Subramanian. Chemical weathering and associated CO_2 consumption in the Godavari river basin, India. *Chem. Geol.*, 264:364–374, 2009.

S. Jobard and M. Dzikowski. Evolution of glacial flow and drainage during the ablation season. *J. Hydrol.*, 330:663–671, 2006.

M. Johannessen and A. Henriksen. Chemistry of snow meltwater: changes in concentration during melting. *Water Resour. Res.*, 14:615–619, 1978.

D. W. Johnson. Base cation distribution and cycling. In D. W. Johnson and S. E. Lindberg, editors, *Atmospheric deposition and forest nutrient cycling*, pages 275–332. Springer Verlag, New York, USA, 1992.

D. Jouvin, P. Louvat, F. Juillot, C. N. Maréchal, and M. F. Benedetti. Zinc isotopic fractionation: why organic matters. *Environ. Sci. Technol.*, 43:5747–5754, 2009.

F. Juillot, C. Maréchal, M. Ponthieu, S. Cacaly, G. Morin, M. Benedetti, J. L. Hazemann, O. Proux, and F. Guyot. Zn isotopic fractionation caused by sorption on goethite and 2-Lines ferrihydrite. *Geochim. Cosmochim. Acta*, 72:4886–4900, 2008.

M. H. A. Jungck, T. Shimamura, and G. W. Lugmair. Ca isotope variations in Allende. *Geochim. Cosmochim. Acta*, 48:2651–2658, 1984.

A. J. Karley and P. J. White. Moving cationic minerals to edible tissues: potassium, magnesium, calcium. *Curr. Opin. Plant Biol.*, 12:291–298, 2009.

S. A. Kasemann, D. N. Schmidt, P. N. Pearson, and C. J. Hawkesworth. Biological and ecological insights into Ca isotopes in planktic foraminifers as a paleotemperature proxy. *Earth Planet. Sci. Lett.*, 271:292–302, 2008.

A. Kaufman Katz, J. P. Glusker, S. A. Beebe, and C. W. Bock. Calcium ion coordination: A comparison with that of beryllium, magnesium, and zinc. *J. Am. Chem. Soc.*, 118:5752–5763, 1996.

M. Kiczka, J. G. Wiederhold, S. M. Kraemer, B. Bourdon, and R. Kretzschmar. Iron isotope fractionation during Fe uptake and translocation in alpine plants. *Environ. Sci. Technol.*, 44: 6144–6150, 2010.

M. Kiczka-Cyriac. *Iron isotope fractionation mechanisms of silicate weathering and iron cycling by plants*. PhD thesis, ETH Zurich, 2010.

J. W. Kirchner. A double paradox in catchment hydrology and geochemistry. *Hydrol. Process.*, 17:871–874, 2003.

C. Kormann. Untersuchungen des Wasserhaushaltes und der Abflussdynamik eines Gletschervorfeldes. Master's thesis, TU Dresden, 2009.

A. Krabbenhöft, J. Fietzke, A. Eisenhauer, V. Liebetrau, F. Böhm, and H. Vollstaedt. Determination of radiogenic and stable strontium isotope ratios ($^{87}Sr/^{86}Sr$; $\delta^{88/86}Sr$) by thermal ionisation mass spectrometry applying an $^{87}Sr/^{84}Sr$ double spike. *J. Anal At. Spectrom.*, 24: 1267–1271, 2009.

P. Krám, J. Hruška, B. S. Wenner, C. T. Driscoll, and C. E. Johnson. The biogeochemistry of basic cations in two forest catchments with contrasting lithology in the Czech Republic. *Biogeochemistry*, 37:173–202, 1997.

K. Kreissig and T. Elliott. Ca isotope fingerprints of early crust-mantle evolution. *Geochim. Cosmochim. Acta*, 69:165–176, 2005.

S. Krishnaswami, J. R. Trivedi, M. M. Sarin, R. Ramesh, and K. K. Sharma. Strontium isotopes and rubidium in the Ganga - Brahmaputra river system: Weathering in the Himalaya, fluxes to the Bay of Bengal and contributions to the evolution of oceanic $^{87}Sr/^{86}Sr$. *Earth Planet. Sci. Lett.*, 109:243–253, 1992.

L. R. Kump, S. L. Brantley, and M. A. Arthur. Chemical weathering, atmospheric CO_2, and climate. *Ann. Rev. Earth Planet. Sci.*, 28:611–67, 2000.

T.P. Labhart. *Geologie der Schweiz*. Ott-Verlag, Thun, 8th edition, 2009.

M. Land and B. Öhlander. Chemical weathering rates, erosion rates and mobility of major and trace elements in a boreal granitic till. *Appl. Geochem.*, 6:435–460, 2000.

E. Landolt and K. M. Urbanska. *Our Alpine Flora*. SAC Verlag, Chur, 2003.

C. H. Langmuir, R. D. Vocke, G. N. Hanson, and S. R. Hart. A general mixing equation with applications to Icelandic basalts. *Earth Planet. Sci. Lett.*, 37:380–392, 1978.

A. C. Lasaga, J. M. Soler, J. Ganor, T. E. Burch, and K. L. Nagy. Chemical weathering rate laws and global geochemical cycles. *Geochim. Cosmochim. Acta*, 58:2361–2386, 1994.

C. H. Lear, H. Elderfield, and P. A. Wilson. Cenozoic deep-sea temperatures and global ice volumes from Mg/Ca in benthic foraminiferal calcite. *Science*, 287:269–272, 2000.

J. Lee, X. Feng, A. M. Faiia, E. S. Posmentier, J. W. Kirchner, R. Osterhuber, and S. Taylor. Isotopic evolution of a seasonal snowcover and its melt by isotopic exchange between liquid water and ice. *Chem. Geol.*, 270:126–134, 2010.

M. Lehning, I. Völksch, D. Gustafsson, T. A. Nguyen, M. Stähli, and M. Zappa. ALPINE3D: a detailed model of mountain surface processes and its application to snow hydrology. *Hydrol. Process.*, 20:2111–2128, 2006.

D. Lemarchand, G. J. Wasserburg, and D. A. Papanastassiou. Rate-controlled calcium isotope fractionation in synthetic calcite. *Geochim. Cosmochim. Acta*, 68:4665–4678, 2004.

E. Lemarchand, J. Schott, and J. Gaillardet. Boron isotopic fractionation related to boron sorption on humic acid and the structure of surface complexes formed. *Geochim. Cosmochim. Acta*, 69: 3519–3533, 2005.

G. E. Likens, F. H. Bormann, N. M. Johnson, D. W. Fisher, and R. S. Pierce. Effects of forest cutting and herbicide treatment on nutrient budgets in the Hubbard Brook watershed-ecosystem. *Ecol. Monogr.*, 40:23–47, 1970.

G. E. Likens, C. T. Driscoll, D. C. Buso, T. G. Siccama, C. E. Johnson, G. M. Lovett, D. F. Ryan, T. Fahey, and W. A. Reiners. The biogeochemistry of potassium at Hubbard Brook. *Biogeochemistry*, 25:61–125, 1994.

G. E. Likens, C. T. Driscoll, D. C. Buso, T. G. Siccama, C. E. Johnson, G. M. Lovett, T. J. Fahey, W. A. Reiners, D. F. Ryan, C. W. Martin, and S. W. Bailey. The biogeochemistry of calcium at Hubbard Brook. *Biogeochemistry*, 41:89–173, 1998.

F. Liu, M. W. Williams, and N. Caine. Source waters and flow paths in an alpine catchment, Colorado Front Range, United States. *Water Resour. Res.*, 40:W09401, 2004.

P. Louvat and C. J. Allègre. Present denudation rates on the island of Réunion determined by river geochemistry: Basalt weathering and mass budget between chemical and mechanical erosions. *Geochim. Cosmochim. Acta*, 61:3645–3669, 1997.

P. Louvat and C. J. Allègre. Riverine erosion rates on Sao Miguel volcanic island, Azores archipelago. *Chem. Geol.*, 148:177–200, 1998.

W. Ludwig, P. Amiotte-Suchet, and J-L. Probst. Enhanced chemical weathering of rocks during the last glacial maximum: a sink for atmospheric CO_2? *Chem. Geol.*, 159:147–161, 1999.

J. Magnusson, D. Farinotti, T. Jonas, and M. Bavay. Quantitative evaluation of different hydrological modelling approaches in a partly glacierized Swiss watershed. *Hydrol. Process.*, 25:2071–2084, 2011.

J. Magnusson, F. Kobierska, S. Huxol, T. Jonas, and J. W. Kirchner. Melt water driven stream and groundwater stage fluctuations on a glacier forefield (Damma gletscher, Switzerland). *Hydrol. Process.*, submitted.

F. Malard, K. Tockner, and J. V. Ward. Shifting dominance of subcatchment water sources and flow paths in a glacial floodplain, Val Roseg, Switzerland. *Arct. Antarct. Alp. Res.*, 31:135–150, 1999.

F. Malard, K. Tockner, and J. V. Ward. Physico-chemical heterogeneity in a glacial riverscape. *Landscape Ecol.*, 15:679–695, 2000.

C. S. Marriott, G. M. Henderson, N. S. Belshaw, and A. W. Tudhope. Temperature dependence of δ^7Li, δ^{44}Ca and Li/Ca during growth of calcium carbonate. *Earth Planet. Sci. Lett.*, 222: 615–624, 2004.

H. Marschner. *Mineral nutrition of higher plants*. Academic Press, London, 1995.

P. Marsh and J. W. Pomeroy. Spatial and temporal variations in snowmelt runoff chemistry, Northwest Territories, Canada. *Water Resour. Res.*, 35:1559–1567, 1999.

B. D. Marshall and D. J. DePaolo. Precise age determinations and petrogenic studies using the K-Ca method. *Geochim. Cosmochim. Acta*, 46:2537–2545, 1982.

B. D. Marshall and D. J. DePaolo. Calcium isotopes in igneous rocks and the origin of granite. *Geochim. Cosmochim. Acta*, 53:917–922, 1989.

M. A. Mast, J. I Drever, and J. Baron. Chemical weathering in the Loch Vale watershed, Rocky Mountain National Park, Colorado. *Water Resour. Res.*, 26:2971–2978, 1990.

C. Mavris, M. Egli, M. Plötze, J. D. Blum, A. Mirabella, D. Giaccai, and W. Haeberli. Initial stages of weathering and soil formation in the Morteratsch proglacial area (Upper Engadine, Switzerland). *Geoderma*, 155:359–371, 2010.

S. B. McLaughlin and R. Wimmer. Calcium physiology and terrestrial ecosystem processes. *New Phytol.*, 142:373–417, 1999.

R. C. Metcalf. The cationic denudation rate of an alpine glacial catchment: Gornergletscher, Switzerland. *Z. Gletscherk. Glacialgeol.*, 22:19–32, 1986.

M. Meybeck. Global chemical weathering of surficial rocks estimated from river dissolved loads. *Am. J. Sci.*, 287:401–428, 1987.

M. Meybeck. Global occurrence of major elements in rivers. In H. D. Holland and K. K. Turekian, editors, *Treatise on Geochemistry*, volume 5: Surface and ground water, weathering and soils. Elsevier, 2003.

C. Mikutta, J. G. Wiederhold, O. A. Cirpka, T. B. Hofstetter, B. Bourdon, and U. Von Gunten. Iron isotope fractionation and atom exchange during sorption of ferrous iron to mineral surfaces. *Geochim. Cosmochim. Acta*, 73:1795–1812, 2009.

W. R. Miller and J. I. Drever. Chemical weathering and related controls on surface water chemistry in the Absaroka Mountains, Wyoming. *Geochim. Cosmochim. Acta*, 41:1693–1702, 1977.

R. Millot, J. Gaillardet, B. Dupré, and C. J. Allègre. The global control of silicate weathering rates and the coupling with physical erosion: new insights from rivers of the Canadian Shield. *Earth Planet. Sci. Lett.*, 196:83–98, 2002.

R. Millot, J. Gaillardet, B. Dupré, and C. J. Allègre. Northern latitude chemical weathering rates: Clues from the Mackenzie river basin, Canada. *Geochim. Cosmochim. Acta*, 67:1305–1329, 2003.

A. C. Mitchell, G. H. Brown, and R. Fuge. Minor and trace element export from a glacierized Alpine headwater catchment (Haut Glacier d'Arolla, Switzerland). *Hydrol. Process.*, 15:3499–3524, 2001.

F. Moynier, S. Pichat, M-L. Pons, D. Fike, V. Balter, and F. Albarède. Isotopic fractionation and transport mechanisms of Zn in plants. *Chem. Geol.*, 267:125–130, 2009.

T. F. Nägler, A. Eisenhauer, A. Müller, C. Hemleben, and J. Kramers. The δ^{44}Ca-temperature calibration on fossil and cultured *Globigerinoides sacculifer*: New tool for reconstruction of past sea surface temperatures. *Geochem. Geophys. Geosyst.*, 1:2000GC000091, 2000.

J. U. Navarrete, D. M. Borrok, M. Viveros, and J. T. Ellzey. Copper isotope fractionation during surface adsorption and intracellular incorporation by bacteria. *Geochim. Cosmochim. Acta*, 75:784–799, 2011a.

J. U. Navarrete, M. Viveros, J. T. Ellzey, and D. M. Borrok. Copper isotope fractionation by desert shrubs. *Appl. Geochem.*, 26:S319–S321, 2011b.

P. Négrel. Geochemical study of a granitic area - the Margeride Mountains, France: Chemical element behavior and $^{87}Sr/^{86}Sr$ constraints. *Aquat. Geochem.*, 5:125–165, 1999.

D. R. Nelson and M. T. McCulloch. Petrogenetic applications of the ^{40}K-^{40}Ca radiogenic decay scheme - A reconnaissance study. *Chem. Geol.*, 79:275–293, 1989.

C. A. Nezat, J. D. Blum, R. D. Yanai, and S. P. Hamburg. A sequential extraction to determine the distribution of apatite in granitoid soil mineral pools with application to weathering at the Hubbard Brook Experimental Forest, NH, USA. *Appl. Geochem.*, 22:2406–2421, 2007.

C. A. Nezat, J. D. Blum, R. D. Yanai, and B. B. Park. Mineral sources of calcium and phosphorus in soils of the northeastern United States. *Soil Sci. Soc. Am. J.*, 72:1786–1794, 2008.

C. A. Nezat, J. D. Blum, and C. T. Driscoll. Patterns of Ca/Sr and $^{87}Sr/^{86}Sr$ variation before and after a whole watershed $CaSiO_3$ addition at the Hubbard Brook Experimental Forest, USA. *Geochim. Cosmochim. Acta*, 74:3129–3142, 2010.

P. W. Nienow, M. Sharp, and I. C. Willis. Velocity-discharge relationships derived from dye tracer experiments in glacial meltwaters: implications for subglacial flow conditions. *Hydrol. Process.*, 10:1411–1426, 1996.

P. Oliva, J. Viers, and B. Dupré. Chemical weathering in granitic environments. *Chem. Geol.*, 202:225–256, 2003.

P. Oliva, B. Dupré, F. Martin, and J. Viers. The role of trace minerals in chemical weathering in a high-elevation granitic watershed (Estibère, France): Chemical and mineralogical evidence. *Geochim. Cosmochim. Acta*, 68:2223–2244, 2004.

P. Ollivier, B. Hamelin, and O. Radakovitch. Seasonal variations of physical and chemical erosion: a three-year survey of the Rhone River, (France). *Geochim. Cosmochim. Acta*, 74:907–927, 2010.

S. Opfergelt, E. S. Eiriksdottir, K. W. Burton, A. Einarsson, C. Siebert, S. R. Gislason, and A. N. Halliday. Quantifying the impact of freshwater diatom productivity on silicon isotopes and silicon fluxes: Lake Myvatn, Iceland. *Earth Planet. Sci. Lett.*, 305:73–82, 2011.

B. D. Page, T. D. Bullen, and M. J. Mitchell. Influences of calcium availability and tree species on Ca isotope fractionation in soil and vegetation. *Biogeochemistry*, 88:1–13, 2008.

M. R. Palmer and J. M. Edmond. Controls over the strontium isotope composition of river water. *Geochim. Cosmochim. Acta*, 56:2099–2111, 1992.

T. Pačes. Sources of acidification in Central Europe estimated from elemental budgets in small basins. *Nature*, 315:31–36, 1985.

M. Pavlov, P. E. M. Siegbahn, and M. Sandström. Hydration of beryllium, magnesium, calcium, and zinc ions using density functional theory. *J. Phys. Chem. A*, 102:219–228, 1998.

S. S. Perakis, D. A. Maguire, T. D. Bullen, K. Cromack, R. H. Waring, and J. R. Boyle. Coupled nitrogen and calcium cycles in forests of the Oregon Coast Range. *Ecosystems*, 9:63–74, 2006.

J. R. Petit, J. Jouzel, D. Raynaud, N. I. Barkov, J-M. Barnola, I. Basile, M. Bender, J. Chappellaz, M. Davis, G. Delaygue, M. Delmotte, V. M. Kotlyakov, M. Legrand, V. Y. Lipenkov, C. Lorius, L. Pépin, C. Ritz, E. Saltzman, and M. Stievenard. Climate and atmospheric history of the past 420,000 years from the Vostok ice core, Antarctica. *Nature*, 399:429–436, 1999.

J. C. Pett-Ridge, L. A. Derry, and J. K. Barrows. Ca/Sr and ^{87}Sr/^{86}Sr ratios as tracers of Ca and Sr cycling in the Rio Icacos watershed, Luquillo Mountains, Puerto Rico. *Chem. Geol.*, 267: 32–45, 2009.

A. Poszwa, E. Dambrine, B. Pollier, and O. Atteia. A comparison between Ca and Sr cycling in forest ecosystems. *Plant and Soil*, 225:299–310, 2000.

P. Raben and W. H. Theakstone. Changes in ionic and oxygen isotopic composition of the snowpack at the glacier Austre Okstindbreen, Norway, 1995. *Nord. Hydrol.*, 29:1–20, 1998.

B. S. Rana, S. P. Singh, and R. P. Singh. Biomass and net primary productivity in cental Himalayan forests along an altitudinal gradient. *Forest Ecol. Manag.*, 27:199–218, 1989.

G. Rauret, J-F. López-Sánchez, A. Sahuquillo, E. Barahona, M. Lachica, A. M. Ure, C. M. Davidson, A. Gomez, D. Lück, J. Bacon, M. Yli-Halla, H. Muntau, and Ph. Quevauviller. Application of a modified BCR sequential extraction (three-step) procedure for the determination of extractable trace metal contents in a sewage sludge amended soil reference material (CRM 483), complemented by a three-year stability study of acetic acid and EDTA extractable metal content. *J. Environ. Monit.*, 2:228–233, 2000.

S. W. Reeder, B. Hitchon, and A. A. Levinson. Hydrogeochemistry of the surface waters of the Mackenzie River drainage basin, Canada - I. factors controlling inorganic composition. *Geochim. Cosmochim. Acta*, 36:825–865, 1972.

L. M. Reynard, G. M. Henderson, and R. E. M. Hedges. Calcium isotope ratios in animal and human bone. *Geochim. Cosmochim. Acta*, 74:3735–3750, 2010.

R. C. Reynolds and N. M. Johnson. Chemical weathering in the temperate glacial environment of the Northern Cascade Mountains. *Geochim. Cosmochim. Acta*, 36:537–554, 1972.

F. M. Richter, D. B. Rowley, and D. J. DePaolo. Sr isotope evolution of seawater: the role of tectonics. *Earth Planet. Sci. Lett.*, 109:11–23, 1992.

C. S. Riebe, J. W. Kirchner, and R. C. Finkel. Erosional and climatic effects on long-term chemical weathering rates in granitic landscapes spanning diverse climate regimes. *Earth Planet. Sci. Lett.*, 224:547–562, 2004.

J. F. Rudge, B. C. Reynolds, and B. Bourdon. The double spike toolbox. *Chem. Geol.*, 265: 420–431, 2009.

W. A. Russell and D. A. Papanastassiou. Calcium isotope fractionation in ion-exchange chromatography. *Anal. Chem.*, 50:1151–1154, 1978.

W. A. Russell, D. A. Papanastassiou, and T. A. Tombrello. Ca isotope fractionation on the Earth and other solar system materials. *Geochim. Cosmochim. Acta*, 42:1075–1090, 1978b.

U. Schaltegger. Unravelling the pre-Mesozoic history of Aar and Gotthard massifs (Central Alps) by isotopic dating - a review. *Schweiz. Miner. Petrog. Mitt.*, 74:41–51, 1994.

E. A. Schauble. Applying stable isotope fractionation theory to new systems. In C. M. Johnson, B. L. Beard, and F. Albarède, editors, *Geochemistry of Non-traditional Stable Isotopes*, volume 55 of *Reviews in Mineralogy and Geochemistry*, pages 65–111. Mineralogical Society of America, Washington D.C., 2004.

A-D. Schmitt, G. Bracke, P. Stille, and B. Kiefel. The calcium isotope composition of modern seawater determined by thermal ionisation mass spectrometry. *Geostandard. Newslett.*, 25: 267–275, 2001.

A-D. Schmitt, F. Chabaux, and P. Stille. The calcium riverine and hydrothermal isotopic fluxes and the oceanic calcium mass balance. *Earth Planet. Sci. Lett.*, 6731:1–16, 2003a.

A-D. Schmitt, P. Stille, and T. Vennemann. Variations of the $^{44}Ca/^{40}Ca$ ratio in seawater during the past 24 million years: Evidence from $\delta^{44}Ca$ and $\delta^{18}O$ values of Miocene phophates. *Geochim. Cosmochim. Acta*, 67:2607–2614, 2003b.

A-D. Schmitt, S. Gangloff, F. Cobert, D. Lemarchand, P. Stille, and F. Chabaux. High performance automated ion chromatography separation for Ca isotope measurements in geological and biological samples. *J. Anal At. Spectrom.*, 24:1089–1097, 2009.

T. Schuler, U. H. Fischer, and G. H. Gudmundsson. Diurnal variability of subglacial drainage conditions as revealed by tracer experiments. *J. Geophys. Res.*, 109:F02008, 2004.

M. Sharp, M. Tranter, G. H. Brown, and M. Skidmore. Rates of chemical denudation and CO_2 drawdown in a glacier-covered alpine catchment. *Geology*, 23:61–64, 1995.

M. Sharp, J. Parkes, B. Cragg, I. J. Fairchild, H. Lamb, and M. Tranter. Widespread bacterial populations at glacier beds and their relationship to rock weathering and carbon cycling. *Geology*, 27:107–110, 1999.

A. M. Shiller. Dissolved trace elements in the Mississippi River: Seasonal, interannual, and decadal variability. *Geochim. Cosmochim. Acta*, 61:4321–4330, 1997.

C. Siebert, T. F. Nägler, and J. D. Kramers. Determination of molybdenum isotope fractionation by double-spike multicollector inductively coupled plasma mass spectrometry. *Geochem. Geophys. Geosyst.*, 2:2000GC000124, 2001.

D. M. Sigman, M. P. Hain, and G. H. Haug. The polar ocean and glacial cycles in atmospheric CO_2 concentration. *Nature*, 466:47–55, 2010.

O. Sigmarsson and S. Steinthórsson. Origin of Icelandic basalts: A review of their petrology and geochemistry. *J. Geodyn.*, 43:87–100, 2007.

N. G. Sime, C. L. De La Rocha, and A. Galy. Negligible temperature dependence of calcium isotope fractionation in 12 species of planktonic foraminifera. *Earth Planet. Sci. Lett.*, 232: 51–66, 2005.

J. I. Simon and D. J. DePaolo. Stable calcium isotopic composition of meteorites and rocky planets. *Earth Planet. Sci. Lett.*, 289:457–466, 2010.

J. I. Simon, D. J. DePaolo, and F. Moynier. Calcium isotope composition of meteorites, Earth and Mars. *Astrophys. J.*, 702:707–715, 2009.

A. Skartveit. Relationships between precipitation chemistry, hydrology, and runoff acidity. *Nor. Hydrol.*, 12:65–80, 1981.

J. Skulan and D. J. DePaolo. Calcium isotope fractionation between soft and mineralized tissues as a monitor of calcium use in vertebrates. *P. Natl. Acad. Sci.*, 96:13709–13713, 1999.

J. Skulan, D. J. DePaolo, and T. L. Owens. Biological control of calcium isotopic abundances in the global calcium cycle. *Geochim. Cosmochim. Acta*, 61:2505–2510, 1997.

J. Skulan, T. Bullen, A. D. Anbar, J. E. Puzas, L. Shackelford, A. LeBlanc, and S. M. Smith. Natural calcium isotopic composition of urine as a marker of bone mineral balance. *Clin. Chem.*, 53:1155–1158, 2007.

R. H. Smittenberg, I. Hajdas, L. Wacker, and S. M. Bernasconi. Soil organic geochemistry and carbon dynamics of an alpine chronosequence. *Geochim. Cosmochim. Acta*, 73:A1242, 2009.

C. Soulsby, J. Petry, M. J. Brewer, S. M. Dunn, B. Ott, and I. A. Malcolm. Identifying and assessing uncertainty in hydrological pathways, a novel approach to end member mixing in a Scottish agricultural catchment. *J. Hydrol.*, 274:109–128, 2003.

K. Stahr, H. W. Zöttl, and F. Hädrich. Transport of trace elements in ecosystems of the Bärhalde watershed in the southern Black Forest. *Soil Sci.*, 130:217–224, 1980.

R. F. Stallard. Tectonic, environmental, and human aspects of weathering and erosion: A global review using a steady-state perspective. *Ann. Rev. Earth Planet. Sci.*, 23:11–39, 1995.

R. F. Stallard and J. M. Edmond. Geochemistry of the Amazon 2. The influence of geology and weathering environment on the dissolved load. *J. Geophys. Res.*, 88:9671–9688, 1983.

R. H. Steiger and E. Jäger. Subcommission on Geochronology: Convention on the use of decay constants in geo- and cosmochronology. *Earth Planet. Sci. Lett.*, 36:359–362, 1977.

T. Steuber and D. Buhl. Calcium-isotope fractionation in selected modern and ancient marine carbonates. *Geochim. Cosmochim. Acta*, 70:5507–5521, 2006.

D. B. Stone and G. K. C. Clarke. In situ measurements of basal water quality and pressure as an indicator of the character of subglacial drainage systems. *Hydrol. Process.*, 10:615–628, 1996.

F. W. E. Strelow. An ion exchange selectivity scale of cations based on equilibrium distribution coefficients. *Anal. Chem*, 32:1185–1188, 1960.

W. Stumm and J. J. Morgan. *Aquatic Chemistry: chemical equilibria and rates in natural waters*. Wiley, 3rd edition, 1996.

E. T. Sundquist and K. Visser. The geologic history of the carbon cycle. In H. D. Holland and K. K. Turekian, editors, *Treatise on Geochemistry*, volume 8: Biogeochemistry. Elsevier, 2003.

A. Taylor and J. D. Blum. Relation between soil age and silicate weathering rates determined from the chemical evolution of a glacial chronosequence. *Geology*, 23:979–982, 1995.

A. B. Taylor and M. A. Velbel. Geochemical mass balances and weathering rates in forested watersheds of the southern Blue Ridge II. Effects of botanical uptake terms. *Geoderma*, pages 29–50, 1991.

S. Taylor, X. Feng, J. W. Kirchner, R. Osterhuber, B. Klaue, and C. E. Renshaw. Isotopic evolution of a seasonal snowpack and its melt. *Water Resour. Res.*, 37:759–769, 2001.

F. Thevenon, F. S. Anselmetti, S. M. Bernasconi, and M. Schwikowski. Mineral dust and elemental black carbon records from an Alpine ice core (Colle Gnifetti glacier) over the last millennium. *J. Geophys. Res.*, 114:D17102, 2009.

E. T. Tipper, M. J. Bickle, A. Galy, A. J. West, C. Pomiès, and H. J. Chapman. The short term climatic sensitivity of carbonate and silicate weathering fluxes: Insight from seasonal variations in river chemistry. *Geochim. Cosmochim. Acta*, 70:2737–2754, 2006a.

E. T. Tipper, A. Galy, and M. J. Bickle. Riverine evidence for a fractionated reservoir of Ca and Mg on the continents: Implications for the oceanic Ca cycle. *Earth Planet. Sci. Lett.*, 247:267–279, 2006b.

E. T. Tipper, A. Galy, and M. J. Bickle. Calcium and magnesium isotope systematics in rivers draining the Himalaya-Tibetan-Plateau region: Lithological or fractionation control? *Geochim. Cosmochim. Acta*, 72:1057–1075, 2008a.

E. T. Tipper, P. Louvat, F. Campas, A. Galy, and J. Gaillardet. Accuracy of stable Mg and Ca isotope data obtained by MC-ICP-MS using the standard addition method. *Chem. Geol.*, 257:65–75, 2008b.

E. T. Tipper, J. Gaillardet, A. Galy, P. Louvat, M. J. Bickle, and F. Campas. Calcium isotope ratios in the world's largest rivers: a constraint on the maximum imbalance of oceanic calcium fluxes. *Global Biogeochem. Cycles*, 24:GB3019, 2010.

M. Tranter. Geochemical weathering in glacial and proglacial environments. In H. D. Holland and K. K. Turekian, editors, *Treatise on Geochemistry*, volume 5: Surface and ground water, weathering and soils. Elsevier, 2003.

M. Tranter, G. H. Brown, A. J. Hodson, and A. M. Gurnell. Hydrochemistry as an indicator of subglacial drainage system structure: a comparison of alpine and sub-polar environments. *Hydrol. Process.*, 10:541–556, 1996.

M. Tranter, P. Huybrechts, G. Munhoven, M. J. Sharp, G. H. Brown, I. W. Jones, A. J. Hodson, R. Hodgkins, and J. L. Wadham. Direct effect of ice sheets on terrestrial bicarbonate, sulphate and base cation fluxes during the last glacial cycle: minimal impact on atmospheric CO_2 concentrations. *Chem. Geol.*, 190:33–44, 2002a.

M. Tranter, M. J. Sharp, H. R. Lamb, G. H. Brown, B. P. Hubbard, and I. C. Willis. Geochemical weathering at the bed of Haut Glacier d'Arolla, Switzerland - a new model. *Hydrol. Process.*, 16:959–993, 2002b.

E. Tresch. Hydrochemistry of the Damma glacier forefield - temporal and spatial variability. Master's thesis, ETH Zurich, http://e-collection.ethbib.ethz.ch/view/eth:30215, 2007.

S. Tsiouris, C. E. Vincent, T. D. Davies, and P. Brimblecombe. The elution of ions through field and laboratory snowpacks. *Ann. Glaciol.*, 7:196–201, 1985.

P. V. Unnikrishna, J. J. McDonnell, and C. Kendall. Isotope variations in a Sierra Nevada snowpack and their relation to meltwater. *J. Hydrol.*, 260:38–57, 2002.

E. Valsami-Jones, K. V. Ragnarsdottir, A. Putnis, D. Bosbach, A. J. Kemp, and G. Cressey. The dissolution of apatite in the presence of aqueous metal cations at pH 2-7. *Chem. Geol.*, 151: 215–233, 1998.

S. C. Van de Geijn and C. M. Petit. Transport of divalent cations: Cation exchange capacity of intact xylem vessels. *Plant Physiol.*, 64:954–958, 1979.

D. Vance, D. A. H. Teagle, and G. L. Foster. Variable Quaternary chemical weathering fluxes and imbalances in marine geochemical budgets. *Nature*, 458:493–496, 2009.

VAW. Gletscherberichte (1881-2002) 'Die Gletscher der Schweizer Alpen', Jahrbücher der Glaziologischen Kommission der Schweizerischen Akademie der Naturwissenschaften (SANW) Versuchsanstalt für Wasserbau, Hydrologie und Glaziologie (VAW) no. 1-122, (http://glaciology.ethz.ch/swiss-glaciers/). ETH Zürich, 2005.

M. A. Velbel. Temperature dependence of silicate weathering in nature: How strong a negative feedback on long-term accumulation of atmospheric CO_2 and global greenhouse warming? *Geology*, 21:1059–1062, 1993.

M. Verbunt, J. Gurtz, K. Jasper, H. Lang, P. Warmerdam, and M. Zappa. The hydrological role of snow and glaciers in alpine river basins and their distributed modeling. *J. Hydrol.*, 282:36–55, 2003.

J. Viers, B. Dupré, M. Polvé, J. Schott, J. L. Dandurand, and J. J. Braun. Chemical weathering in the drainage basin of a tropical watershed (Nsimi-Zoetele site, Cameroon): comparison between organic-poor and organic-rich waters. *Chem. Geol.*, 140:181–206, 1997.

J. Viers, B. Dupré, J. J. Braun, S. Deberdt, B. Angeletti, J. Ndam Ngoupayou, and A. Michard. Major and trace element abundances, and strontium isotopes in the Nyong basin rivers (Cameroon): constraints on chemical weathering processes and elements transport meachanisms in humid tropical environments. *Chem. Geol.*, 169:211–241, 2000.

J. Viers, P. Oliva, A. Nonell, A. Gélabert, J. E. Sonke, R. Freydier, R. Gainville, and B. Dupré. Evidence of Zn isotopic fractionation in a soil-plant system of a pristine tropical watershed (Nsimi, Cameroon). *Chem. Geol.*, 239:124–137, 2007.

N. Vigier, K. W. Burton, S. R. Gislason, N. W. Rogers, S. Duchene, L. Thomas, E. Hodge, and B. Schaefer. The relationship between riverine U-series disequilibria and erosion rates in a basaltic terrain. *Earth Planet. Sci. Lett.*, 249:258 – 273, 2006.

D. Viviroli and R. Weingartner. The hydrological significance of mountains: from regional to global scale. *Hydrol. Earth Syst. Sci.*, 8:1016–1029, 2004.

J. L. Wadham, A. J. Hodson, M. Tranter, and J. A. Dowdeswell. The rate of chemical weathering beneath a quiescent, surge-type, polythermal-based glacier, southern Spitsbergen, Svalbard. *Ann. Glac.*, 24:27–31, 1997.

J. C. G. Walker, P. B. Hays, and J. F. Kasting. A negative feedback mechanism for the long term stabilization of Earth's surface temperature. *J. Geophys. Res.*, 86:9776–9782, 1981.

J. M. Watkins, D. J. DePaolo, C. Huber, and F. J. Ryerson. Liquid composition-dependence of calcium isotope fractionation during diffusion in molten silicates. *Geochim. Cosmochim. Acta*, 73:7341–7359, 2009.

D. J. Weiss, T. F. D. Mason, F. J. Zhao, G. J. D. Kirk, B. J. Coles, and M. S. A. Horstwood. Isotopic discrimination of zinc in higher plants. *New Phytol.*, 165:703–710, 2005.

L. R. Welp, J. T. Randerson, J. C. Finlay, S. P. Davydov, G. M. Zimova, A. I. Davydova, and S. A. Zimov. A high-resolution time series of oxygen isotopes from the Kolyma River: implications for the seasonal dynamics of discharge and basin-scale water use. *Geophys. Res. Lett.*, 32: L14401, 2005.

A. J. West, A. Galy, and M. Bickle. Tectonic and climatic controls on silicate weathering. *Earth Planet. Sci. Lett.*, 235:211–228, 2005.

A. F. White and A. E. Blum. Effects of climate on chemical weathering in watersheds. *Geochim. Cosmochim. Acta*, 59:1729–1747, 1995.

A. F. White and S. L. Brantley. The effect of time on the weathering of silicate minerals: why do weathering rates differ in the larboratory and field? *Chem. Geol.*, 202:479–506, 2003.

A. F. White, A. E. Blum, M. Schulz, T. D. Bullen, J. W. Harden, and M. L. Peterson. Chemical weathering rates of a soil chronosequence on granitic alluvium: I. Quantification of mineralogical and surface area changes and calculation of primary silicate reaction rates. *Geochim. Cosmochim. Acta*, 60:2533–2550, 1996.

A. F. White, T. D. Bullen, V. Vivit, M. S. Schulz, and D. W. Clow. The role of disseminated calcite in the chemical weathering of granitoid rocks. *Geochim. Cosmochim. Acta*, 63:1939–1953, 1999a.

A. F. White, A. E. Blum, T. D. Bullen, D. V. Vivit, M. Schulz, and J. Fitzpatrick. The effect of temperature on experimental and natural chemical weathering rates of granitoid rocks. *Geochim. Cosmochim. Acta*, 63:3277–3291, 1999b.

P. J. White and M. R. Broadley. Calcium in plants. *Ann. Bot.*, 92:487–511, 2003.

P. J. White, J. Banfield, and M. Diaz. Unidirectional Ca^{2+} fluxes in roots of rye (*Secale cereale* L.). A comparison of excised roots with roots of intact plants. *J. Exp. Bot.*, 43:1061–1074, 1992.

J. G. Wiederhold, N. Teutsch, S. M. Kraemer, A. N. Halliday, and R. Kretzschmar. Iron isotope fractionation in oxic soils by mineral weathering and podzolization. *Geochim. Cosmochim. Acta*, 71:5821–5833, 2007.

B. A. Wiegand, O. A. Chadwick, P. M. Vitousek, and J. L. Wooden. Ca cycling and isotopic fluxes in forested ecosystems in Hawaii. *Geophys. Res. Lett.*, 32:L11404, 2005.

M. E. Wieser, D. Buhl, C. Bouman, and J. Schwieters. High precision calcium isotope ratio measurements using a magnetic sector multiple collector inductively coupled plasma mass spectrometer. *J. Anal At. Spectrom.*, 19:844–851, 2004.

M. W. Williams and J. M. Melack. Solute chemistry of snowmelt and runoff in an alpine basin, Sierra Nevada. *Water Resour. Res.*, 27:1575–1588, 1991.

F. Wombacher, A. Eisenhauer, A. Heuser, and S. Weyer. Separation of Mg, Ca and Fe from geological reference materials for stable isotope ratio analyses by MC-ICP-MS and double-spike TIMS. *J. Anal At. Spectrom.*, 24:627–636, 2009.

A. Wong, A. P. Howes, R. Dupree, and M. E. Smith. Natural abundance ^{43}Ca NMR study of calcium-containing organic solids: A model study for Ca-binding biomaterials. *Chem. Phys. Lett.*, 427:201–205, 2006.

WRB. *World Reference Base for Soil Resources 2006 - A framework for International Classification, Correlation and Communication.* Number 103. Food and Agriculture Organization of the United Nations, Rome, Italy, 2006.

C. Yang, K. Telmer, and J. Veizer. Chemical dynamics of the "St. Lawrence" riverine system: δD_{H_2O}, $\delta^{18}O_{H_2O}$, $\delta^{13}C_{DIC}$, $\delta^{34}S_{sulfate}$, and dissolved $^{87}Sr/^{86}Sr$. *Geochim. Cosmochim. Acta*, 60:851–866, 1996.

J. C. Yde, N. T. Knudsen, and O. B. Nielsen. Glacier hydrochemistry, solute provenance, and chemical denudation at a surge-type glacier in Kuannersuit Kuussuat, Disko Island, West Greenland. *J. Hydrol.*, 300:172–187, 2005.

J. Zachos, M. Pagani, L. Sloan, E. Thomas, and K. Billups. Trends, rhythms, and aberrations in global climate 65 Ma to present. *Science*, 292:686–693, 2001.

E. A. Zakharova, O. S. Pokrovsky, B. Dupré, and M. B. Zaslavskaya. Chemical weathering of silicate rocks in Aldan Shield and Baikal Uplift: insights from long-term seasonal measurements of solute fluxes in rivers. *Chem. Geol.*, 214:223–248, 2005.

P. Zhu and J. D. Macdougall. Calcium isotopes in the marine environment and the oceanic calcium cycle. *Geochim. Cosmochim. Acta*, 62:1691–1698, 1998.

Acknowledgements

First of all I would like to thank my supervisors for giving me a large degree of independence in my work and the freedom to pursue my own research objectives. My PhD work was conducted as a joint venture between the Isotope Geochemistry group under Bernard Bourdon and the Soil Chemistry Group under Ruben Kretzschmar. Although I had four supervisors, I have found it incredibly useful to always receive a range of opinions from different viewpoints, which luckily never contradicted each other, and substantially enriched my work. I thank Ben Reynolds for introducing me to column chemistry, the Triton, setting up the double-spike method and his patience in helping me to improve my writing; Jan Wiederhold for always promptly responding to any problems I had; Ruben and Bernard for giving me the opportunity to do this project and the valuable discussions regarding data interpretation. The discussions with Bernard on his trips back to Zurich from Lyon were particularly helpful. In addition to my supervisors, I wish to thank Stephan Kraemer and James Kirchner for agreeing to co-examine this thesis.

I am indebted to Manu Lemarchand and Ed Tipper for thought-provoking discussions, advice and encouragement over the last few years. I also thank Ed for his patience with the endless drafts of the water paper. It has been a great pleasure to work with both of them and I learned a lot from them. Mirjam Kiczka and Greg de Souza both helped immensely with their knowledge of the Damma and isotope fractionation processes and provided many of the field samples analysed in this thesis.

I have been very fortunate to have had the opportunity to interact with people from a wide range of disciplines through the BigLink project. I thank the whole BigLink team for the soil sampling trip, maintaining the infrastructure in the fieldsite, support and interesting discussions, particularly Stefano Bernasconi for overseeing the whole project; Jan Magnusson for help with all hydrology related problems; Daniel Farinotti for help with glacier dynamics and organising the winter sampling trip, probably the best sampling trip; Monika Welc for teaching me about mycorrhiza and determining the degree of root infection; and Hans Göransson for teaching me about alpine plants. The water sampling trips would not have been possible without all the people who helped filter and carry samples: Ben, Jan W, Manu, Ed, Hans, Rienk Smittenberg, Mirjam, Giuditta Fellin, Rasmus Thiede, Merle Gierga and Christoph Burkhardt. I thank Martin Theiler for help during a very cold diurnal sampling campaign and Thomas Reichmuth for help capturing a 'rain event'. During my time at ETH I was lucky enough to fulfil my dream of doing fieldwork in Greenland. I thank Jemma Wadham for letting me join the expedition and everyone who was in the field for that month, especially Jörg Rickli for company during the endless filtration of water samples.

Measuring does not always go to plan and I am grateful for the help of Andreas

Stracke, Ben and Felix Oberli for help troubleshooting problems on the Triton and to Mathieu Touboul, Ulrik Hans, Antoine Roth and Sarah Aciego for discussions on measurement and loading techniques. I also thank Felix for his immense effort in running the labs; Kurt Barmettler for running the soil chemistry labs and help with XRF, ICP-OES and FAAS measurements; Irene Ivanov-Bucher for teaching me how to transform a piece of rock into mineral separates; Iso Christl for help setting up and measuring anion concentrations; Maria Coray Strasser for measuring oxygen isotopes; Marie-Thèrese Bär for showing me how to make and degass filaments; Martin Imseng for help with developing the sequential extraction method; Felix Oberli, Urs Menet, Andreas Süsli, Donat Niederer, Heiri Baur and Colin Maden for building/maintaining/repairing lab equipment and the mass specs; Bruno Rütsche and Colin for IT support, and Britt Meyer and Valentina Müller-Weckerle for administrative support.

The working atmosphere at ETH had been immensely enjoyable and inspiring and for that I would like to thank everyone in the Isotope Geochemistry and Soil Chemistry groups. In particular, I want to thank to my office mates past and present who made the office a fun place to work: Giuditta, Christoph, Antoine and Jörg.

PhD study requires a strong antidote and I would like to thank everyone who accompanied me into the mountains at the weekends to ski, climb, walk and forget about work. I am grateful to the support of my family despite my Dad believing that the last four years have been more 'play' than 'work'. I would like to thank Jakob for translating the thesis summary and for getting me to Switzerland and of course, finally, to Gjermund for providing the motivation to keep writing and actually get this thesis finished. Nå er det ferdig!

Die VDM Verlagsservicegesellschaft sucht für wissenschaftliche Verlage abgeschlossene und herausragende

Dissertationen, Habilitationen, Diplomarbeiten, Master Theses, Magisterarbeiten usw.

für die kostenlose Publikation als Fachbuch.

Sie verfügen über eine Arbeit, die hohen inhaltlichen und formalen Ansprüchen genügt, und haben Interesse an einer honorarvergüteten Publikation?

Dann senden Sie bitte erste Informationen über sich und Ihre Arbeit per Email an *info@vdm-vsg.de*.

Sie erhalten kurzfristig unser Feedback!

VDM Verlagsservicegesellschaft mbH
Dudweiler Landstr. 99
D - 66123 Saarbrücken
www.vdm-vsg.de

Telefon +49 681 3720 174
Fax +49 681 3720 1749

Die VDM Verlagsservicegesellschaft mbH vertritt

Printed by Books on Demand GmbH, Norderstedt / Germany